Praise for *T*

ABOUT THE AUTHORS

Lydia V. Pyne is a visiting fellow at Drexel University's Pennoni Honors College. She has an MA and PhD in the history and philosophy of science, as well as an MA in anthropology. She has done extensive fieldwork in archaeology and paleoanthropology in Arizona, New Mexico, South Africa, Ethiopia, and Uzbekistan.

Stephen J. Pyne, Lydia's father, is a Regents Professor in the School of Life Sciences, Arizona State University. An award-winning environmental historian, he is the author of *Voyager, Year of the Fires, The Ice*, and *How the Canyon Became Grand*, among many other books.

Lydia V. Pyne and Stephen J. Pyne

THE LAST LOST WORLD

Ice Ages, Human Origins, and the Invention of the Pleistocene

PENGUIN BOOKS

PENGUIN BOOKS
Published by the Penguin Group
Penguin Group (USA) Inc., 375 Hudson Street,
New York, New York 10014, U.S.A.

Ⓟ

USA | Canada | UK | Ireland | Australia | New Zealand | India | South Africa | China
Penguin Books Ltd, Registered Offices: 80 Strand, London WC2R 0RL, England
For more information about the Penguin Group visit penguin.com

First published in the United States of America by Viking Penguin,
a member of Penguin Group (USA) Inc. 2012
Published in Penguin Books 2013

THE LIBRARY OF CONGRESS HAS CATALOGED THE HARDCOVER EDITION AS FOLLOWS:
Pyne, Lydia V.
The last lost world : ice ages, human origins, and the invention of the Pleistocene /
Lydia V. Pyne and Stephen J. Pyne.
p. cm.
Includes bibliographical references.
ISBN 978-0-670-02363-9 (hc.)
ISBN 978-0-14-312342-2 (pbk.)
1. Geology, Stratigraphic—Pleistocene.   2. Paleogeography—Pleistocene.
I. Pyne, Stephen J., 1949–   II. Title.
QE697.P96 2012
551.7'92—dc23   2011038464

Printed in the United States of America
1   3   5   7   9   10   8   6   4   2

Set in Columbus MT Std
Designed by Francesca Belanger

For Stan

—LVP

For Sonja, both past and future, as always

—SJP

What a piece of work is a man, how noble in reason, how infinite in faculties, in form and moving how express and admirable, in action how like an angel, in apprehension how like a god: the beauty of the world, the paragon of animals! And yet to me what is this quintessence of dust?

—William Shakespeare, *Hamlet,* Act 2, Scene 2

I don't want to belong to any club that will accept me as a member.

—Groucho Marx

Contents

Table of Figures

Prologue

Mossel Bay, South Africa

I PUT MY CHOPSTICKS and tweezers down and looked out of the cave's entrance. I saw the sun glinting off the ocean's surface and dolphins cutting through the water as they splashed around offshore. After six weeks in this cave, tucked off the coast of Mossel Bay, I was so accustomed to the pounding surf against the rocks that the roar outside of the cave proper seemed downright insignificant. Although coastal South Africa was one of the more idyllically situated field projects I had worked on during grad school, my mind wasn't on the shrubby fynbos outside the cave's entrance. My attention was very narrowly focused on the square of sediment I had been tweezering away at for weeks. I turned back to the stratigraphic lens and repositioned the chopsticks to get a better angle to continue paring away the sediment.

It took my eyes a few minutes to refocus from the bright light outside to the soft glow of the fluorescent bulb hanging above me. The cave was deep enough in the coastal bedrock that sunlight barely illuminated the entrance. Those of us in the back of the cave were excavating our precisely measured units courtesy of artificial lights strung along the walls. Several lights were in fixed positions shining on the sides of the squares' vertical stratigraphic profiles, and a couple of bulbs hung freely on wires so excavators could more closely examine artifacts and changes in soil color.

As a grad student, my job was to excavate, bag artifacts, and keep careful records from the field season—every bone, stone, and stratum was documented, photographed, and filed away for future analysis. Ultimately, we wanted to be able to explain the cave as a whole, its artifacts, and its setting within the context of the Pleistocene of South Africa. Our inquiry into the broader natural history of South Africa was nothing new. Since the 1770s, coastal South Africa has had intellectual appeal as a place for researchers to come and collect plants, animals, and artifacts and to take these collections back to various museums and universities for study. Carl Linnaeus's students Carl Peter Thunberg and Anders Sparrman were quick to publish their findings about the biodiversity of the South African fynbos and early twentieth-century archaeologists were intrigued with the cave sites and shell middens found along the coast. Each set of researchers in South Africa has tried to understand South Africa's natural history by working out how their research fits into a broader intellectual schema.

In our field season, we were no different. We were rehearsing, like our intellectual predecessors in natural history, a constant dialectic— an intellectual tension, if you will—trying to balance making sense of a holistic system by understanding its parts. To be able to do this, we needed to take the cave apart according to its different inorganic stratigraphic pieces and see what parts the cave was made up of— soils, artifacts, and their positions relative to each other. The parallel between us and our intellectual history was practically complete if you could swap out a quaint cabinet of curiosities for the hundreds of sandbags protecting the cave's surface from trampling feet, the soil samples piling up along the walls, and the day's artifacts bagged, tagged, cataloged, and carefully placed in plastic bins to be stored at the local museum.

However, I came to my interest in the Pleistocene and archaeology from a very nontraditional path. With a background in history, my sense for explanation focused heavily on narrative—what story or stories exist, how they are told, and what constitutes a reasonable telling. The original draw to archaeology had been the appeal of hands-on history. However, archaeology and, indeed, many other sciences that are Pleistocene-focused work to make sense of the Pleistocene through reductionism: by taking the complex phenomenon of human history and analyzing of it through its smallest components, whether those parts are isotopes or artifacts. There's a certain necessity in this approach to the Pleistocene—and even a certain parsimony—from the perspective of a researcher, but it leaves little room for broader themes or complex explanations. From my perspective as someone also trained in the humanities, the Pleistocene I knew as a field archaeologist left little place for the questions, debates, and intellectual tensions that have piqued people for millennia and were the themes of other scholarship.

However, this begs the question of what a humanities-informed treatment of the Pleistocene might look like and what kind of scholarship might work to inform it. As a geologic epoch, the Pleistocene lasted from approximately 2.6 million to 10,000 to 12,000 years ago. It includes an ice age, a shakeout of seas and lands, Earth's fifth great extinction event, and the appearance and evolution of the hominins. It's a world of mammoths and woolly rhinos, erectines and Neanderthals, ice sheets and warm spells; it's a time of extinctions and origins.

However, the Pleistocene is also an idea. And as an idea, the Pleistocene appeared in the early nineteenth century, inseparable from other concepts and assumptions of those times. It then underwent intellectual upheavals and an evolution of its own, as profound as the inundations of ice and pluvials. From it emerged a new origins story, whose *fiat lux* was announced in the secularized tropes of modern science. Inherited explanations from Aristotle and Aquinas died out as newly crafted ones by Charles Darwin and Ernst Haeckel shouldered their way in. The sciences split from the humanities, leaving a peculiar parallax of understanding by which to look on the past.

The Pleistocene, in brief, has a dual character both as a geologic time and as a cultural idea. The one is a hard history of rock, ice, bone, and artifacts while the other is a soft history of words, images, and concepts. Both epoch and idea have a story to tell, and the latter, a story to tell about how the story itself is told. Together they describe a place in the geography of imagination and another of geologic time whose difference helps define our own time and selves. They create what Paul Martin once called the "last lost world."[1] They describe the world that made our modern one.

This book project grew out of my dual interests in the Pleistocene and the humanities. It seeks to move beyond the questions of "right" and "wrong" interpretations of the Pleistocene and works to situate the questions of the epoch into a broader intellectual schema and to show that debates, arguments, people, and fossils fit into age-old tropes and behave in ways that stand outside of simple explanations found within respective scientific disciplines. It makes a case for pragmatism and pluralism—that there are many ways to understand the Pleistocene. We offer one that focuses on broadly humanist themes, which we intend to complement, not compete with, prevailing Pleistocene studies. It would seem that we follow Plato and want to join that fluorescently lit reductionism deep in the cave with the light of the outside world.

—Lydia Pyne

PART ONE

HOW THE PLEISTOCENE GOT ITS ICE

Glacial epochs are great things, but they are vague—vague.

—Mark Twain, *Life on the Mississippi* (1883)

Such is the unity of all history that anyone who endeavours to tell a piece of it must feel that his first sentence tears a seamless web.

—Frederic Maitland, "Prologue to a History of English Law" (1898)

Mark Twain was right, but what makes the composite of glacial epochs known as the Pleistocene vague is also what makes it great. Among geologic epochs it has sought to embrace two unique events, each with a distinctive narrative and both at odds with the principles by which the rest of the geological timescale is ordered. They identify a break, a rift, not only in geologic time but in how that chronology is understood. Out of the Pleistocene's vague greatness they tear the seamless web of history.

The first event is the sudden upheaval in global climate that manifested itself most spectacularly in continent-scale ice sheets and their geographic analogs in the form of massive lakes and deserts. The second event is the evolution of the modern hominins, whose tenure on Earth has, with equal force, both defined and confused the boundaries of the epoch. Both events are processes that translate into narratives. The first, or geographic, narrative tells how the epoch got its ice. The second, or hominin, narrative tells how it got its most distinctive creature. Both topics trace out, and then acting together like shears of a scissors, cut out the borders of the Pleistocene.

But more than simple natural events both ice and hominins are also ideas. That's what transfigures data points into story, adds cultural value, and creates understanding. Those ideas did not emerge by spontaneous generation out of the mud of artifacts and sediment cores; they have their own historical settings, their own intellectual lineages, their own separate and collective narratives. They describe how the epoch acquired meaning. They tell how, as it were, the Pleistocene got its mind.

Ice, humans, ideas—the Last Lost World is the outcome of all three, flowing into and out of each other like braided channels on an outwash plain. In our own retelling we look to the sciences to identify the natural processes at work and to craft a basic chronology of events. But we look to other scholarships to comment on how such information becomes narrative and how an understanding of the Pleistocene plenum, including the evolution of hominins, has itself evolved.

Chapter 1

Rift

Most of us have a hankering for the good things on both sides of the line.

—William James, *Pragmatism* (1907)

LIKE ALL STORIES, this one begins with a Maitland-like tear in the seamless webs of nature and culture. Geologically, it involves a rip in the deep crust of Africa that marks a shift in tempo and events, and culturally, the tear refers to a parting in ways of understanding, a chasm between modes and purposes of explanation. Over time— almost three million years for nature, less than three thousand for Western civilization—the rips have widened into a rift, and that rift is not only where the story originates but is in some ways itself the story.

African Rift

Its Great Rift Valley marks Africa in much the same way that flanking mountains do the other continents. But where the Andes, or the Sierra Cascades, or the Great Dividing Range have resulted from collisions of lands forcibly joined, the Rift Valley is the outcome of tension from Africa splitting, literally tearing the continent in two.

The process commenced when, some seven million years ago, a mantle-driven hot spot beneath the crust caused much of the continent to bulge upward; rips appeared in the crust along a loose line from Ethiopia to Mozambique; and gradually these separate cracks connected. Over time they began joining, to become one of the defining features of Earth: an immense Rift Valley, or what early observers termed the Great Wall of Africa. The collapsed crest was the outcome of regional uplift, like the center of an arch falling as its supporting ends swayed outward. The resulting ravine, here narrow, there broad, alternately filled with lava and lakes. In places the crustal tears ran in parallel, collapsing the land between into huge depressions; Lake Victoria formed in one such basin. Sentinel volcanoes guarded the flanks. Then some two to three million years ago, the hot spot moved and the rending ceased. What began as a series of rips and cuts ended as a crustal scar. The sundering had waxed and then waned, like the passing of a tectonic tide.[1]

What remained was the Great Rift. It was corridor, habitat, cauldron, and refugium. It both fissioned and fused—an obstacle and an opportunity. As a geologic marker it was equally adamant and ambiguous. There was no denying its geographic presence; it was to Africa what the Rockies were to North America or the Alps to Europe. It accented the asymmetry of Africa east and west, as much as that continent's bulging top and narrowing bottom did north and south. Valley and wall broke the continuity of climate across the continent; the lands east of the Rift fell under the rhythms of the Asian monsoon. Its rise and fall through history was murkier, much less distinct than its sheer-walled flanks unless viewed through the telescoping lens of geologic epochs. Yet here, too, it was both a bridge and a barrier, not for land but for time, and through time, for story.

As Earth time goes, the changes came suddenly. What the geological timescale calls the Miocene had been unusually warm and wet globally, a robust epoch that had virtually boiled over with species. Around 5.3 million years ago, the landscape of the Miocene began to slow, as Earth seemed to pause for breath. Land, sea, air, and life—the matrix of natural history reorganized into what became the Pliocene epoch, as each element, both the rocky and the fluid, underwent a series of suturings and sunderings. The African Rift was one of them, but so were the continued elevation of the Himalayas and the Andes, which created similar breaks in continental climate, and the fusion of North and South America at the Isthmus of Panama, which forced a general realignment of planetary climate.

Together such reconfigurations of land, sea, and air helped splinter the roughly homogenized climate of the Miocene, which became more regional, more seasonal, and overall much drier and cooler. So too the continental fissurings shook up lush biota, splitting off pieces like lithic flakes from a core. The evolving matrix of bridges and barriers shut out, let in, and mixed up. Land spans and rifts acted like funnels and filters, pouring one flora and fauna into another while sieving out others. Species followed such emerging landscapes, particularly the savannas, competed with immigrants and émigrés as old lands opened up, and sought out newly distinctive niches. The flow of species across former frontiers sparked profound upheavals that led equally to profuse colonizations and a body count of extinctions. In the Great American Interchange across Panama, llamas and tapirs went south and armadillos and porcupines north, while carnivorous opossums and terror birds died out. All this allowed the biotic overspillage of the Miocene to sort itself out and regroup over the course of three million years.

As the Pliocene wound down, the sluggish grating of rocks yielded to the fast flowing of water and air—climatic factors that hastened, deepened, and sharpened. The change was foreshadowed by a bout of glaciation that coincided roughly with the cessation of the Rift. Soon global climate conditions supplanted mantle heating as a prime mover of planetary change. When that happened the Pliocene, acting as a kind of historical rift, segued into the Pleistocene. The warm-wet exuberance of the Miocene gave way to the cool-dry violence of the Pleistocene, in which the frost-thaw of glaciation subjected biotas to relentless rhythms of firing and quenching. What the Pliocene had sifted, the Pleistocene then shook, warmed, froze, and sent forth into the modern world.

Renaissance Rift

Such is the Pleistocene as a geologic moment—its material text, so to speak. Like artifacts in a museum, however, its evidence and relics need organization into exhibits, captions, and guidebooks if they are to fit into a larger schema. Since the epoch involves historical change, people turn to narratives to explain it, and to glosses on those narratives. Both will vary by times and temperaments.

Formal learning came and went in Western civilization amid flickering climates of opinion as baffling in their causes and as profound in their consequences as those that governed the ice ages. Framing the terms of modern understanding began with the Renaissance, a revival of learning based on texts written in ancient times, ones lost and then recovered. The former challenged the prevailing scholarship, which was organized around theology. Since the new inquiry studied the material world, it was labeled secular; since it focused on people and what it meant to be human, it became broadly known as humanism; and since it engendered a rebirth of learning, it was proudly called a renaissance. It provided a second gloss to that proffered by theology and its scholars.

The process by which ancient learning was recovered occurred between the twelfth and sixteenth centuries. It sparked revivals in various places as circumstances favored, then migrated and started again. By the sixteenth and seventeenth centuries it was mutating into another form of inquiry altogether, one whose practitioners read the book of nature, not simply the texts of scripture and antiquity. By the late eighteenth century this scientific revolution had turned its attention to such curiosities as bones, buried fossils, and strata that did not seem to come from the contemporary world. By the early nineteenth century the understood age of the Earth had grown to such an extent that it required periodization, and acquired a discipline of its own. The epoch of Earth time closest to the present day came to be called Pleistocene. Over the next two centuries the boundaries of this epoch were stretched and reorganized to suit novel data, fresh styles of expression, and updated temperaments. The onset of modern science struck scholarship as the onset of ice struck Earth.

While the Pleistocene as an idea drew the borders for the Pleistocene as a parcel of geologic time, it was not apparent to everyone how its pieces might connect internally or how the epoch joined to past and present. The debates over the Pleistocene might stand as an exemplar for what William James, an observer of the exfoliating controversies, described as a general quarrel between two temperaments that he described as "tough-minded" and "tender-minded." The tough-minded tend toward the material, the skeptical, and the empirical. The tender-minded veer toward the idealistic, the dogmatic, and the rationalist in the sense of seeking out informing reasons, or principles.

In James's day the fissure between tough and tender tracked the shifting borders between science and religion, neither of which he found satisfactory by itself: "You find an empirical philosophy that is not religious enough, and a religious philosophy that is not empirical enough for your purpose." Most philosophers, James continued, sought out the best of both approaches. "Facts are good, of course—give us lots of facts. Principles are good—give us plenty of principles. The world is indubitably one if you look at it in one way, but as indubitably is it many, if you look at it in another. It is both one and many—let us," he urged, "adopt a sort of pluralistic monism." After all, he noted, with Emersonian disregard for logical consistency, most of us want the best of both.[2]

Substitute "humanities" for "religion" and the widening chasm between temperaments and the world they imagine could hold today as well. To the tough-minded, the Pleistocene appears as a plexus of events, of processes acting on earth, air, water, and life, measured by meters and minutes, all contained within a chronological frame. In their valuation, other notions have no heft: they belong in a lost world of murky metaphysics and misty imaginings. To the tender-minded, the epoch is a notion, perhaps even a sentiment, measured by figures of speech, and whose data become comprehensible by art, philosophy, and narrative that provide a necessary commentary on the hard text of stone and chronicle. Their narratives are unstable—the rims that frame the rift move, after all. As tough-minded research struggles to incorporate more data and the unsettling presence of hominins, the chronological borders of the epoch shift. For the tender-minded the arc of meaning relocates with changes in perception, modes of expression, and the gossamer fabric of social meanings. The enduring narrative— the one they share—is how they interact across the rift.

Past Imperfect

Like any part of the past, the Pleistocene is difficult to know. Its material records are fragments of stone and bone, caches of sediment and charcoal, daubs of air trapped in soil and ice. They are data points, like a scatter of dots. How to connect them by abstractions such as theories, models, informing ideas, and organizing conceits is both simple and not obvious because the possible arrangements, or figurations, are many, even if bounded by a neuro-Kantian hard-wiring of understanding. But where the dots deal with time, the connections will take the loose form of a narrative. They will tell a story.

Stories about the past are easy to invent and hard to verify. Paradoxically, the more distant the past, the simpler the task, because the various patterns that one can imagine proliferate; the space-time cone of narrative possibilities expands. As more becomes known, however, the stories become more complicated. From time to time they must be redesigned or else risk disappearing under a scree field of information. The Miocene and Pliocene, which together precede the Pleistocene by twenty million years, seem homogenous because we know relatively little about them and they are usefully remote in time; the texture of their history blurs, like the blue haze around a distant mountain. By contrast, the Holocene, which succeeded the Pleistocene and covers the past 10,000 to 12,000 years, seems so crowded with artifacts and records we can hardly see its mountains for the pebbles on the road.

A narrative, moreover, has its own internal logic. While each observer of the Pleistocene will tell the story according to his own vantage point, disciplinary inclination, and purpose, an appeal to narrative imposes structural demands of its own. In particular it requires special attention to beginnings and endings, for they set the anchor points for the bridging arc that must span them. In truth, these relationships are reciprocal. If the endpoints set the story line, it is equally true that the narrative arc determines where the anchor points must reside. Together they decide whether the story is ironic, providential, tragic, redemptive, coherent, convincing, or ultimately explanatory.

For example, begin with a land bridge such as Beringia, and the story moves one way. Begin with a barrier like the flooded Red Sea, and it trends another. Open with a glaciation or an interglacial—the story line leads to different conclusions. View evolution from a naturalistic perspective, and *Homo sapiens* is not a progressive end point of evolution; geography and genomes might have yielded many alternatives (and did). View those events from the perspective of narrative, and *Homo sapiens* could only have happened one way, for only one sequence could have produced the species at its exact conclusion. The paradox is that many narratives are possible, each according to a different trait or theme, but each must follow its theme from origin to terminus in a predetermined arc.

This logic applies to any narrative, whether of natural or human history. The power of the explanation derives precisely because it restricts its scope; it tears the seamless web of time; it blocks out of the scene those events and objects that do not align with its theme; it brings closure to the text. Part of its validity lies with its elegance, the power of its artistry. But the need to restrict is true equally for science, which has progressed exactly because it has confined itself to certain kinds of propositions and evidence and has tossed everything else aside. Both science and art necessarily leave out stuff, and the tendency for each is to ultimately leave out the other's approach altogether.

From their origins these two narratives—call them the geographic and the genomic—have formed a tag team that has withstood all challengers. The most recent epochs of Earth time have been a story of us and the circumstances that have made us what we are.

Yet both themes unsettle a simple story of the Pleistocene. A reliance on climate unsettles because it depends on the natural sciences, which constantly introduce new findings and theories and thus subvert any particular narrative. Rather than use data to tell stories, scientists tell stories about how they got their data. A reliance on humanity forces us to consider those features of human life that, while critical, are not amenable to sciences altogether. By absorbing a creature that, in the Great Chain of Being, lies between ape and angel, it appeals to a scholarship between, as it were, the sciences and the humanities. As an idea the Pleistocene is thus doubly unstable. What endures beyond particular geographic or genomic narrative is the process of narrating.

For those keen to make the subject a positivist science, one solution is to scrap narrative altogether, which leaves the formal economy of knowledge to thrive in a black market of popular enthusiasms. Others accept that narrative will crowd into the tent, whether invited or not, and appeal to context to help contain its potentially destabilizing wobbles. Comparisons, contrasts, continuities real or imagined—all can shrink the distance between thematic and aesthetic closure. By appealing to epochs before and after, by comparing hominins to coresiding creatures such as mammoths, cave bears, and horses, by contrasting philosophies of knowing, a usable narrative can emerge.

And if, as a result, an element of distortion persists, a foreshortening of time and detail that hedges into exaggeration, some hyperbole seems more than appropriate, for the Pleistocene has always loomed larger than life; and by most evidence, it *was* an era ripe for exaggeration—of big events, of big creatures, of big upheavals, of big consequences. A quickening of climatic tides, that variously flooded, drained, dried, and glazed in ice vast regions; not once, but repeatedly. Oceans rose and fell. Continents swelled and shrank, as their offshore shelves emerged and submerged beneath seas, and as islands and mainlands joined and split off. Flora and fauna wandered widely—colonizing, retreating, hiding, advancing; splintering, and reorganizing, bubbling up from evolutionary swamps, disappearing over evolutionary cliffs—a grand era of extinctions. It was an age of geomorphic gigantism as well, with colossal landforms no less than *Megatheria* or Shasta ground sloths.

Then, 10,000 to 12,000 years ago, the Pleistocene inflected into the Holocene, an epoch so increasingly dominated by a genus of hominins, modern humanity, that observers have proposed naming its most recent phase the Anthropocene. Of all the planetary changes wrought by the Pleistocene, the emergence of this creature may be the most significant. At the height of the Pleistocene, perhaps a dozen or more species of hominins flourished on three continents. When it ended, there was one hominin, *Homo sapiens*, who had expanded into every niche reachable from the African landmass and was following the retreating ice sheets and pluvial lakes and all the new lands shining under the great warmth that had succeeded the big chill. That colonization would continue into every nook and cranny of Earth, and even beyond. By the twenty-first century the climate that had powered this revolution was itself being altered by the magnitude of anthropogenic tinkering, and the creature responsible for this upheaval demanded an explanation.

Some stories are more interesting than others, some less so; some more useful, some less. As narratives they all require only endpoints and a thematic arc by which to span them, and they will succeed to the extent that they agree with what the audience determines is suitable evidence, reasoning, and relating, or with what William James considered our ultimate logic, "our more or less dumb sense of what life honestly and deeply means."[3]

Chapter 2

Ice

The long summer was over. For ages a tropical climate had prevailed over a great part of the earth, and animals whose home is now beneath the Equator roamed over the world from the far South to the very borders of the Arctics. . . . But their reign was over. A sudden intense winter, that was also to last for ages, fell upon our globe.

—Louis Agassiz, on the coming of the Ice Age (1866)

IN 1837 LOUIS AGASSIZ, a Swiss naturalist, borrowed the expression *Eiszeit* ("Ice Age") from a friend and gave the near landscape of geologic time a Romantic cast that still infuses it. As more discoveries poured in, it appeared the Pleistocene had opened with a fanfare of cold that spawned a hellish brood of ice sheets, dune deserts, and swollen lakes. The consequences, as Agassiz exulted with melancholy fervor, were severe. "The appearance of this great cover of ice must have brought with it the extinction of all organic life on the surface of the globe," he gushed. "To the movement of a powerful creation succeeded the silence of death."[1]

"Ice Age" collapses into a simplistic phrase the complex character of a glacial epoch that was not a singular event—a frozen Noachian flood—but a jostling of earth, water, air, and life. It was a vast syncopation of geologic, atmospheric, oceanic, and biotic rhythms. The ice, and all the aftershocks of an ice inundation, appeared and vanished. The tidal flows of climate waxed and waned to longer mixed beats that pumped, stored, and realigned planetary waters on a vast scale and rewired the circulation of the atmosphere. Collectively these pulses sent the living world into a fast march, demanding that species readjust not only to the reorganized ensemble of land, sea, and air, but to one another, to all those other organisms now crowding together, or unleashed over newly defined landscapes, or joined in common transit.

It is not possible to isolate a solitary cause—a primary driver—from which all the other effects follow like a steam engine, powering gears and pulleys, each in necessary sequence, since every twist and tweak affected all the others. Each bout of glaciation, broadly recognized, reset the starting conditions for the next. The planet did not revert from each nominal cycle to its prior state like a piston returning to top dead center after firing. Rather, it more resembled a kaleidoscope whose churning patterns looked ever similar but never identical. As each grand rhythm ceased, Earth was irreversibly different from what it had been at the start; and so it was with the era overall. Compared to times before and after, the epoch gradually chilled, despite spikes of intense warming.

If the Pleistocene lacks a singular cause, it had in its character a singular vision—ice. Later it added hominins as a defining feature, but it kept ice or ice surrogates, such as pluvial lakes, and ice-wrought effects, like a lowered sea level, as the epoch's essence, for the ice was not merely a passive product of the evolving matrix but an active agent in its own right. It made the Pleistocene both unique and compelling.

What most astonished its discoverers, however, was the apparent rapidity of the ice's arrival. Not only did the climate abruptly leap outside normal expectations, but it did so with the velocity of a falling star. A similar sense of suddenness infected its intellectual shape, for if the Pleistocene had one matrix of matter, it had another of mind.

The appreciation of what had happened in the past was itself made possible by the advent of the upheaval of thought known as modern science. Its advent had a longer gestation, one remarked upon and recorded for several centuries; but its tempo had steadily quickened, like a meteor dropping under the accelerating pull of gravity, and by the time Agassiz announced his discoveries, the inherited canon was being smothered by new data, archaic disciplines were disappearing like dew, and terms like "natural philosophy" were yielding to more modern labels such as "science." Yet it seemed to come as abruptly as Agassiz's global winter. In Herbert Butterfield's memorable meditation on the origins of modern science, "Similarly, though everything comes from antecedents and mediations—and these may always be traced farther and farther back without the mind ever coming to rest—still, we can speak of certain epochs of crucial transition, when the subterranean movements come above ground, and new things are palpably born, and the very face of the earth can be seen to be changing." Science quickly covered the terrain of scholarship as relentlessly as Pleistocene ice had the landscapes of Europe.[2]

The contributing causes to the Ice Age were many. The ice came because of larger arrangements of continents and oceans; because of realigned currents of sea and air; because of snow catchments near the poles, on mountains, and shallow upper-latitude basins; because sunlight strengthened and weakened with the wobbles and waverings of Milankovitch cycles; because Earth had been long cooling and now passed a tipping point. So, too, the conception of an Ice Age had many sources, though from cultural rather than climatic drivers. It happened because of large reshufflings among hard and soft disciplines; because of the secular advance of modern science; because a Little Ice Age affected Europe at a time when naturalists were available to take notice; and because of catalytic, even charismatic, personalities. Arguments and ideas came and went like ice sheets and pluvial lakes and left behind a complex intellectual terrain of conceptual moraines and wave-cut terraces to be colonized by what became contemporary thought.

The fact remains that the Pleistocene no more carries an inherent meaning than does a fossil mandible teased out of shale. Provenance and setting is all: the bone takes its meaning from its context. So, too, does the Pleistocene, whose context depends on a history of thought, not simply on ice-plowed soils and greenhouse gases and runoff. The narrative arc of the Pleistocene begins with ice and ends with idea.

Ice Age

When the Pleistocene commenced, part of its elemental matrix was fixed and part was in flux. The fixed part was the solid Earth. Only 2.6 million years long, the epoch lacked sufficient time to witness major tectonic realignments. Volcanism, flooding, and land sculpture went on as always, some locally pronounced; but Earth did not re-lay its tectonic tiles in any major way during the epoch.

The flux part was literally its fluids—the gases that made up its atmosphere, the waters that filled its rivers, lakes, and oceans. What most characterized the Pleistocene geologically was that its liquids repeatedly crystallized into solids in the form of ice sheets that plated over seas, capped mountains, and mounded above a significant fraction of the Earth's land surface. They created an icy, more plastic tectonics atop the plate tectonics of the crust. These ice sheets absorbed much of the planet's waters, and acting like a suddenly uplifted plateau or ridge, they rerouted the old flow of air and waters; in the northern hemisphere they assumed the dimensions of new continents. In the twinkling of a geologic eye, they did what might have required untold eons of slow tectonic shuffling and shearing. For life on the surface, the effect was to foreshorten geologic time. What would have taken millions of years happened in thousands.

Such sudden changes in the Earth's geography forced even more abrupt adjustments. What the ice did not claim directly, it affected indirectly; it could touch what it couldn't grasp. It reorganized the physical geography of the planet, and piling up amid Earth's biota, it triggered mass migrations, reorganizations, and eradications. The ice did not, moreover, happen once and leave the Earth to adjust, like an asteroidal impact. Even more rapidly than it appeared, it departed. Then the cycle began again, and again, perhaps fifty times throughout the Pleistocene.

The Ice Cometh and Goeth

Only four times in Earth history has ice defined a geologic era. There was a long glaciation in the late Precambrian, between 800 million and 600 million years ago. An Ordovician-Silurian glaciation occurred from 460 million to 420 million years ago, possibly followed by a sharp episode during the Devonian. A sustained outbreak occurred in Permo-Carboniferous times, 320 million to 250 million years ago. The modern glaciation commenced 2 million to 3 million years ago. All resulted from favorable geographies of land and ocean and, on land, of suitable traps and catchments for snow. Once that planetary matrix was established, other factors determined how extensively and how frequently the ice came and went. In reality, each ice epoch was a composite of many glaciations.[3]

What all shared was an arrangement between Earth and Sun that determined how much sunlight reached the planet, and on Earth, an ordering of lands and seas that decided how that heat got moved around. Snow had to accumulate; the buildup could set in motion positive feedbacks that gathered still more snow; and massed ice could then resculpt the master matrix. On a planetary scale it meant that heat collected at the tropics could not move freely to the poles, which left some places cold out of season. Just this happened when, at the end of the Pliocene, the Bering Strait opened, the Mediterranean flooded and drained, and the Central America Cordillera fused at Panama to shut off the flow of subtropical waters. Earth acquired a macrogeography primed for a glacial outbreak.

This was not the only possible arrangement for an ice age. Each of the great glacials (and perhaps others yet undiscovered) has had its own distinctive matrix, and there were additional singularities that entered into the equation that could override astronomical cycles. One is surely the evolution of life. The Precambrian ice age lasted the longest and occurred before life colonized the continents. The Permo-Carboniferous frosted over stretches of a supercontinent, Pangaea, during a time of floral abundance and the mass burial of organics. Surely the ramifying character of life as a regulator of greenhouse gases and albedo (surface reflectivity) affected the dynamics of those events. The Pleistocene glaciation introduced yet another singularity, the evolution of hominins, who began to tinker with the internal dynamics of planetary heating and made this epoch like no other.

Despite Agassiz's formative vision, this latest Ice Age was not a global glazing. The ice flourished only in select lands, and on those sites it waxed and waned; there were regulators that governed the machinery on shorter timescales. Current theories attribute this effect to changes in how Earth alters its orientation relative to the Sun, rhythms now known as Milankovitch cycles after the Serbian mathematician, Milutin Milankovitch, who first elaborated them.

Three cycles—or "orbital forcings"—are most relevant. One stretches the distance between the Earth and the Sun, one tilts the Earth toward and away from the Sun, and one wobbles the Earth's axis. The first describes the way the Earth moves around the Sun. The ellipse that traces its orbital pathway lengthens and compresses, a change called eccentricity. At times the Earth reaches farther out, its orbit becoming more ovoid, and then retracts into something closer to a circle. This stretching occurs on a roughly 100,000-year cycle, and perhaps one at 412,000 years. A second cycle refers to the degree by which the Earth's axis veers away from a vertical plane; that is, the degree to which its equator tilts (its obliquity) from the plane of its orbit (the ecliptic). Today that tilt measures at 23.4°; in the past it has swung to 21.8°. The tilt rocks back and forth approximately every 41,000 years. The third cycle is a wobble in the axis, like that of a spinning top as it slows. The precession of the equinoxes (or simply, the precession) inscribes a 22,000-year rhythm, or more precisely, a major precession of 23,700 years and a minor one of 19,000. Collectively the three cycles act as a global thermostat.[4]

Stretching, tilting, wobbling—the combinations of their interactions are endless and maddeningly complex when translated into impacts on mountains, seas, and continental shelves. Milankovitch believed the ideal circumstances for glaciation required minimum obliquity, relatively high eccentricity, and an aphelion that coincided with a northern hemisphere summer, or in other words, a tiny tilting, a lot of stretching, and a wobbling that put Earth facing away from the Sun during the summer. Such arrangements make it possible for snow to come and stay where most of the continental landmasses are. Filtering this enormous cacophony of rhythms to detect a pattern is akin to sifting a cave stuffed with sediment to find a fossil molar. Yet it happens: insolation signatures do poke through the debris.[5]

Once started, the expanding snow's albedo can leverage a weak signal into a shout. A tiny change can trigger huge ones, as a small switch can turn on a giant air-conditioning unit. Glacial episodes resemble asset bubbles in this way. Once begun, the tendency is to add still more, and once halted, to shed rapidly. An increase in ice encourages more ice, while a reduction pushes the system into still further and faster reductions. The buildup is slow; the descent, a crash. Even vast ice sheets can dissolve in several thousand years, which is astonishing when one considers the many and complicated lag times required in removing ice. The overall pattern is one of long stretches of cold glaciation broken by sudden, fleeting interglacials. Some 80 percent of the Pleistocene was glacial, and no interglacial lasted more than 12,000 years.

The current glacial epoch commenced shortly after the completed Isthmus of Panama forced a realignment of ocean circulations. The first tremors of glaciation appeared around 2.6 million years ago, although local terrain determined how that climatic shock was manifest. In central Siberia it meant permafrost; in the Alps, glaciers; on the Canadian Shield, an ice sheet about two and a half miles thick; in Australia, a vast gyre of sand dunes; across much of the northern hemisphere, immense plains of loess downwind from ice masses; in the African Rift, overflowing lakes; and in North America's Great Basin, a pulsing of playas and inland salt seas. Beating to its own internal pacemakers, this climatic oscillation has continued unchecked until modern times.

Ice Earth

While the material matrix had its hard and soft parts, ice was the geologic glue that held them together. What made the matrix mobile was the fact that the ice could not only come and go but, as water, transmute from solid to fluid, and could force even seemingly rigid features to bend and buckle. In short, Earth made ice, and then ice remade Earth.

The solid matrix proved in reality to be malleable. As glaciation transferred water from oceans to continents, it unburdened the deep seas and weighed down the land. The sheer load of the ice caused the crust to slowly depress, like a stone placed on paraffin. The lands along the fringe bulged up, further sculpting the tedious outward flow of the ice. Slowly the Earth adjusted, sinking or rebounding as the ice mounds came and went, working out a new balance with a rigid crust sandwiched between the plastic ice and a plastic mantle—a process of accommodation called isostasy. Since the ice could move much faster than the crust, equilibrium always lagged, and the surface warped and flexed over thousands of years.

The Laurentide ice sheet caused Hudson Bay to sink about 1,400 feet; the rebound, uplifting slowly century by century, has risen some 1,000 feet over the last 7,500 years. The Fenno-Scandinavian ice sheet warped the Baltic down about 3,000 feet; its shorelines are still rising, perhaps three feet per century, as waters drain away to leave skerries that become marshlands, and eventually firm land. The Antarctic ice sheet bowed the continent some 4,000 feet down, pushing much of it beneath sea level and leaving the entire Earth vaguely pear-shaped. On smaller scales, ice scooped out basins; planed gorges; rumpled terrain with moraine, loess, and sand; and froze soils. On a larger order, it caused oceans to rise and fall; it exposed and submerged continental shelves; and it reconfigured bridges and barriers, forcing exotic species to mingle and old biotas to pry apart.[6]

As glaciation sponged up and released waters, it directly affected oceans, lakes, rivers, subsurface acquifers, and the humidity of the air.

It removed water from the seas and put it, crystallized, on land. It upset the circulation of the world ocean, deflecting, for example, the path of the Gulf Stream.

At the height of glaciation the global sea level dropped nearly four hundred feet. The sunken shoreline created a new dispensation of continental geographies. The continents grew, although less space was habitable. They were also joined in ways impossible before. Beringia bonded Asia and North America. The Isthmus of Panama linked North and South America. Ireland and Britain became continuous with Europe. The Indonesian archipelago welded to southeast Asia; New Guinea and Tasmania fused to Australia; between them the Lombok Strait narrowed and shortened to a depth of sixty-six feet. It was now possible to walk from the Cape of Good Hope to Cape Horn, and with a short raft ride over the Wallace Line, from Gibraltar to Tasmania.

Ice ages on the continents expressed themselves not only as glacials but as so-called pluvials. These were times of exceptional wetness. Where outlets remained open, the enhanced rainfall poured through existing channels, entrenching rivers and down-cut gorges, much as glaciers did. But they also took the form of lakes, a watery clone of the ice sheets. Some such lake sheets occupied permanent sinks, including the depressions of the African Rift Valley; the interior depressions that sagged behind the Himalayan, Elburz, and Caucasus mountains; that penned in the Great Basin of the American West; and that rumpled northern Africa in long-wave swales. Such sites deepened, widened, and even overflowed under the rains of a cooling climate. The Great Salt Lake swelled into an ancient Lake Bonneville that washed over a third of Utah. Lake Agassiz submerged over 135,000 square miles, four times the size of Lake Superior today. Africa's Lake Chad overflowed to 115,000 square miles. Asia's Aral and Caspian seas joined to make a monstrous subcontinental body of water over 400,000 square miles. Like rising and falling oceans, the sloshing lake levels led to stacks of wave-cut terraces, and like ice masses, these hefty interior seas forced the crust to flex and buckle beneath them. But much of the water was as fleeting as hot money.

During glacials they added to ice sheets; during interglacials they melted into proglacial lakes, often prone to eruptive flooding, or they carved gorges, or poured over fast-receding falls. The breaching of an ice dam allowed Lake Missoula to scour out the channeled scablands of Washington. The Laurentide ice sheet left a watery echo of itself in the soggy landscape of the Canadian Shield and that arc of freshwater lakes along the Shield's edge, altogether almost 10 percent of the Earth's fresh water.[7]

Eventually, the ice sheets shrank to watery lees. They endured as much wizened lakes, quirky streams, cobbled-together rivers, sodden or frozen soils, boggy lowlands, pond-pocked plains, and the delightfully named "deranged drainage" of the Shield, the watery equivalent to loess and dune fields. Those waters follow contours set by an icy terrain now vanished, like a tree bent by winds that no longer blow. They join in freakish confederations, or meander through channels far too vast for their present flows, or shrivel into oases, or sink into sands as buried aquifers or creeping, underground rivers. Some of the adjustments, like Niagara Falls, have thousands of years yet to run, the fluvial equivalent of isostatic rebound after ice mounds have melted.

Compared to crust and ocean, the atmosphere is thin, and its gases react far more quickly than earth or water. It is with a chill in the air that glaciations begin, and once begun the informing ice acts to deepen the cooling and reroute the general circulation. So interwoven is Earth's circulation of air that what happens in one place will connect with others, even if the linkages might seem to belong to a Rube Goldberg machine whose inputs can appear far removed from its outputs.[8]

Once an ice age has begun, the atmosphere both transmits and tempers its effects. An ice sheet can block airflow as surely as a new mountain range; an iced-over ocean or subcontinent can rewire the connections by which the planet is heated and cooled; glacial and periglacial features can make and deflect local winds, nudge or shove storm systems into new tracks, and trim or stretch seasons. During the Pleistocene they heightened the Asian monsoon. They deflected the polar front that traces the boundary between warm and cold currents in the North Atlantic. They rejigged the El Niño–Southern Oscillation. They affected winds that stirred outwash plains and dune fields and spread dust that could hang suspended for months, and they moved dirt from Tibet to Greenland and from the Sahara to Amazonia.

The glacials also changed the chemistry of the atmosphere, which in turn affected the dynamics of heating and cooling that powered glaciations. Of particular interest is carbon. A goodly fraction of planetary carbon has been stored in rock such as limestone or been buried as organic matter that gradually lithified into coal or petroleum; so, too, methane and carbon have been trapped into the crystal lattices of clathrates and stashed on the ocean floor. The most mobile, however, is lodged in Earth's biota, which has sequestered and liberated it by photosynthesis and respiration, and by living and dying. With the mutuality typical of life, by such means the biosphere altered the workings of the atmosphere.[9]

The living world, in brief, is not a passive recipient of geographic change but an active agent in it, and a quickening one. In some places ice simply obliterated everything not ice, and elsewhere it changed the environment in ways that forced organisms to relocate. But as life moved and reorganized, it stored and released carbon, and by changing the atmosphere Earth's biota altered the climate that nominally drove the living world.

The process of filtering and mixing the Miocene legacy was already well advanced thanks to the Pliocene's geographic filtering. The fast flickering of climate, with its lagging ice sheets and unsteady seas, acted like a biotic bellows to force and hasten migration, extinction, and rebirth. Within a few thousand years (perhaps as few as twenty to thirty generations of the larger mammals), habitats underwent cycles of freezing and thawing, seasonal migration routes became choked off or suddenly unblocked, gene pools were drained and refilled, throngs of plants and animals found themselves cleaved apart, flung widely or crowded together, and then reaggregated.

Genera and biomes advanced and receded. The Pleistocene became one of geologic time's great epochs for evolutionary upheaval, the fifth of known episodes of mass species creation and obliteration. In the end, renewal never quite matched in numbers gained what extinction took away.

Species could move, and did, both pulled and pushed by external and internal factors. When conditions deteriorated, as ice, tundra, dunes, or lakes submerged or remade landscapes, species retired to places of ecological sanctuary, or refugia. These were islands isolated by rising seas, or mountain peaks that served as islands, amid expansive ice or dune fields; they were wet gulleys or oases amid deepening deserts; they were caves or otherwise sheltered sites where species could weather the climatic storms. Other sanctuaries were simply more favorable settings; here refugees of broken biotas retreated and regrouped at lower (or higher) elevations, or beyond the reach of periglacial winds and chill. Such sites were biotic reserves where those organisms that had survived the culling crowded together. A few scattered species escaped beyond those throngs, the biotic equivalent of erratic boulders. When ice released its grip, colonization became less a matter of close-packed ecosystems marching systematically outward than of a stream of refugees or a land rush, a helter-skelter sprawl that would take millennia to sort out amid an inconstancy of conditions that never quite settled down during the Pleistocene. Generalists had better survival rates than specialists; the small, better than the large.

The Pleistocene thus had both places of refuge and sites of unrest; it had times of calm and times of turmoil; it had bridges and barriers and a climatic colander whose filters species had to pass through. It was a long trek paced by stasis and stress. One theory asserts that the stasis, particularly the calms of the refugia, is what allowed species to diversify. Another points instead to stress, the relentless turnover of habitats and ceaseless wanderlust, to account for the record of species spawned and killed off. Overall, the climatic whipsawing left behind a rough-hewn land and a biota tempered for the nimble, the quick, and the opportunistic.[10]

Unlike the rhythmic ice, which returned more or less without regard to prior conditions, the biotic narrative is one that arcs, pulse by pulse, warmth followed by chill, from a time of plenty to one of scarcity. The Pleistocene story ends much differently than it began.

Eiszeit and Zeitgeist

The intellectual narrative of the Pleistocene also follows a long arc that ends far from where it began. It has its defining feature—modern science rather than ice—which acted uniquely on the inherited landscape. Along the way it spins and pulses, cycle after cycle, with ideas often exerting a reach beyond their immediate grasp.

A signature moment in that history is 1642, the year Galileo Galilei died and Isaac Newton was born. Together the two events trace the chronology of the scientific revolution from conception to exemplar. What had been shuffling along for centuries, with a sharpened tempo since Nicolaus Copernicus published De Revolutionibus in 1543, now slammed into place. It was a cultural event akin to the final fusion of the Isthmus of Panama for natural history. Old forms of argument, logic, and rhetoric now acted on the revamped matrix of scholarship with startling power.

What emerged was modern science. Not only was this "a new factor," as Herbert Butterfield observed, "but it proved to be so capable of growth, and so many-sided in its operations, that it consciously assumed a directing rôle from the very first, and, so to speak, began to take control of the other factors—just as Christianity in the middle ages had come to preside over everything else, percolating into every corner of life and thought." It proceeded to push other modes of inquiry to the side, like so much moraine. It deflected and rerouted the circulation of information. Under its impact, old ideas went extinct and new ones arrived. With a positive feedback that allowed it to aggregate more and more, modern science progressively defined the character of the epoch. It proclaimed the cultural world that made ours.[11]

As they reread the book of nature, scholars who embraced the new learning saw with fresh insight. Exposed strata, assemblages of rocks, fossil seashells, mastodon bones—all had existed forever, like Milankovitch cycles. Ordinary people who stumbled upon them attributed them to vanished giants or monsters and sports of nature. Intellectuals, hunched over texts, ignored them. But as modern science spread, those objects found a new setting with which to interact. By the eighteenth century, naturalists were prowling the countryside seeking out such oddities for cabinets of curiosities, and struggled to interpret them within the new dispensation. In 1735 Linnaeus published *Systema Naturae*, an organizing schema that did for natural history what Newton's *Principia Mathematica* and *Opticks* had done for natural philosophy. There was now, at least in principle, a place to put such artifacts.

There was also a time. In 1648 Bishop James Ussher, assiduously tabulating the genealogical record of the Bible, published a preliminary chronology of Earth, dating its origin some 6,000 years before the present. Three centuries later, in 1948, when the 18th International Geological Congress redeclared the boundaries of the Pleistocene, the known age of the Earth had expanded nearly a millionfold, from 6,000 years to 4.6 billion. The task fell to geology, an emerging science, to map and name this new world of time as cartographers were doing for the new worlds unveiled by voyages of geographic discovery.

Still, Louis Agassiz's proclamation of an *Eiszeit* was not simply another artifact to be absorbed within the enveloping medium of geologic time but an informing event that ordered an age. Certainly it helped structure an era in the history of geology—and of cultural enthusiasms generally. Here was a Romantic vision in an age that looked to Nature for inspiration as well as enlightenment. Only a year before Agassiz's announcement, Ralph Waldo Emerson had published his manifesto of transcendentalism, *Nature*, and Thomas Cole had painted *The Oxbow*, inaugurating a passion for landscape art; two years later the U.S. Army organized a Corps of Topographic Engineers to explore the exotic lands of the far west. An Ice Age was Nature's past as a New World. Unsurprisingly, Agassiz found America a congenial place and immigrated there to establish at Harvard a Museum of Comparative Zoology. The Eiszeit had met the *Zeitgeist*.

Science and sentiment merged. Within thirty years, however irritating and distasteful to the sensibilities of the old establishment, the vision of an Ice Age had achieved consensus, overwhelming the emerging concept of uniformitarian time as its ice did the Val de Bagnes in the Swiss Alps. By 1841 Edward Forbes could write Agassiz that his ideas had made "all the geologists glacier-mad here, and they are turning Great Britain into an ice house."[12] But Forbes was an early convert and champion. Geology as a discipline was itself still in its adolescence. Most of Earth lay unexplored by naturalists and known lands had yet to be examined for evidence of past glaciation; the residual ice sheets on Greenland and Antarctica were known only at their fringe. It took thirty years before Archibald Geikie could provide convincing data for Britain, and another thirty before James Geikie could expand that survey over a whole Earth. It required almost another century to track down its causes.

A Time Whose Idea Had Come

The organization of deep time was the special mission of geology, and it set about chronicling rocks into a temporal stratigraphy. Oddly, perhaps, the fresher stuff was harder to fit into the design, and was left to coat the surface like sprinkles on a cake. In 1756 Johann Lehmann identified two broad categories of rocks: an older (Primary) and younger (Secondary), overlain by recent soils. Four years later Giovanni Arduino proposed to enlarge the newer layers into a Tertiary era. In 1783, the term "geology" was coined, the known age of Earth began to lengthen, subdivisions started to proliferate, and the search for organizing principles commenced in earnest. Eventually they found a more universal chronometer in the fossils of sedimentary strata. In the early nineteenth century fossils were the future of the past.[13]

If the past proved harder to know, the present was harder to classify. In 1829 Jules Desnoyers proposed to split off more contemporary times into a Quaternary era. The breakthrough, however, came when Charles Lyell created a more detailed geochronology in his three-volume *Principles of Geology* (1830–33). Using a meticulous record of forty thousand mollusk fossils excavated from the Paris Basin by Gerard Paul Deshayes, Lyell subdivided the Tertiary age into four phases. Since the absolute ages of the rocks (and their embedded fossils) were unknown, Lyell back-sited from the present. He called the oldest epoch the Eocene, deriving the term from the Greek *eos* (dawn or earliest) and *kainos* (recent); the next oldest, the Miocene, from the Greek *meion* (less); then the Pliocene, from *pleion* (more), divided into Older and Newer. Beyond that, leading to the present, was the Post-Tertiary, a term he preferred to Quaternary. Of the index fossil mollusks, the Eocene had only 3.5 percent identical to those of the present day; the Miocene, 17 percent; the Pliocene, 35 percent to 50 percent. Closer to the present the proportions increased: the Post-Pliocene had 90 percent to 95 percent.[14]

It was the Post-Tertiary that struggled to find its footing. Even the name was conflated, as Lyell mixed the Latin *post* with the Greek *kainos*. Lyell divided it into two phases: the Post-Pliocene and the Recent, and of course Post-Pliocene immediately got confused with Post-Tertiary. He ignored Quaternary, a term he never accepted. The Recent addressed the age "tenanted by man," which at the time barely extended beyond the chronicles of the Bible. In Lyell's hands, the geologic record was thus consistent in its criteria and continuous in its sequencing.

That quickly changed. Within thirty years Lyell would himself write *The Geological Evidences of the Antiquity of Man*, extending human history into Earth history. Barely had the ink dried on the *Principles* when Agassiz issued his dramatic annunciation of an Ice Age. In 1839 Lyell replaced Newer Pliocene with Pleistocene, advancing the Greek from the comparative to the superlative (*pleistos*, most recent). He quickly regretted the decision and withdrew it, not wishing to distinguish too closely the near from the now. But Edward Forbes reinstated the term in 1846 as a synonym for the Ice Age, and it has never left, despite Lyell's reluctance and stubborn continuance of his original term in later texts. Likewise, Lyell's use of the Recent was challenged in 1867 by Paul Gervais's Holocene (from the Greek *holos*, meaning whole. Why is unclear; perhaps Gervais had in mind another Greek expression, "synecdoche," meaning a part for the whole, all of which does not explain the origins of the terms but may help account for the extinction of the classics as a part of the modern science curriculum). Together, Pleistocene and Holocene made up a composite era, the Quaternary, which fully replaced Post-Tertiary, completing Lyell's linguistic rout. In 1873, Lyell accepted Forbes's redefinition; two years later he died, and resistance ended. In 1885 Holocene was formally submitted to the International Geological Congress and became the preferred working term.[15]

The issue went beyond squabbling over what names to give the thickening geologic past. It went to the question of how that past should be organized. Those who pondered geology whole wanted the present to be continuous with the past—that was the essence of uniformitarianism. They wanted the Quaternary to obey the same rules as the other subdivisions. But for many practitioners two new principles seemed to define the beginning and the end of the Pleistocene. One was the Ice Age, which would identify the onset of the epoch. The other was the "tenancy of man," which would announce its terminus. The Holocene would embrace the age of humanity as the Pleistocene did the age of ice.

But this left matters unsettled and ambiguous, because glaciation was not simultaneously global, the tenancy of humanity kept being pushed back to the origins of that glaciation epoch, and there was no clear-cut inflection from Pleistocene to Holocene, since there was no evidence that the cycling of glacials and interglacials had suddenly ceased around 10,000 years ago. Instead the epoch became an anomaly, because it was defined by climatic boundaries, recorded through ice and the relict geomorphic "fossils" of glaciation, rather than by extinctions recorded in fossils of organic evolution. The evolution of life was idiographic, and directional, which made it a better index for long chronicles than climate, which might recycle like the seasons. How orbital geometries equated with mollusk habitat was unclear. The Quaternary became the problem child of the geological timescale.

So even as the Pleistocene, as a term, became fixed, its borders as a geologic epoch remained elastic, its duration unknown, and its age only relative. In 1863, four years after Darwin published *On the Origin of Species*, Lyell estimated the duration of the Pleistocene at 800,000 years. In 1900 W. J. Sollas put it at 400,000 years. In 1909, based on Alpine terracing (wrongly, as it happens), Albrecht Penck and Eduard Brückner dated the epoch at 650,000 years. As a rule of thumb, working geologists bracketed the epoch as "the last million years." In 1948 an International Geological Congress combined the glacial with the paleontological indices and identified the lower boundary of the Pleistocene with the advent of cold-loving species in the Mediterranean Basin, some 1.64 million years ago, an event recorded in the Vrica stratum in Calabria, southern Italy, which marked "the horizon of the first indication of climatic deterioration in the Italian Neogene succession," a conclusion affirmed by the 7th International Union for Quaternary Research Congress in 1965 and again in 1983. Chronological creep nudged the origin to around 1.8 million years. Interestingly, the official report on the 1948 Congress noted, however, that the session on the Plio-Pleistocene boundary was "mainly of interest to specialists." Granted that organisms had begun colonizing the continents around 420 million years ago, the Pleistocene constituted less than one-half of one percent of the estimated duration of terrestrial life.[16]

Adjusting the boundaries of the geologic timescale was nothing new; this is how scientists parsed and analyzed the field. What irked geologists generally, however, was the insistence by Quaternarists for unique criteria, which seemed a case of special pleading. They wanted to bring the quirky Pleistocene, with its apparitions of ice floods and wandering hominins, to heel. Meanwhile, a revolution engulfed the earth sciences in the 1960s and 1970s, typically summed up as "plate tectonics," which created novel methods for dating, fresh sources of data, and a chronicle not stored among the fickle fossils of the continents. The dating methods were radiometric, paleomagnetic, and chemical, in the form of assorted isotopes preserved in rocks, ice, and oozes. The critical data came from the deep oceans, not from lands scoured by ice; it derived from paleomagnetic surveys of lava from spreading centers, from deep-ocean sediment cores (particularly those rich in fossil foraminifera), from the analysis of relic isotopes. Flickering on and off with almost digital precision, such methods promised a more universal chronology based on physical and chemical measurements not bound by the depositional, erosional, and biotic quirks of specific landscapes that had been subjected to endless surficial scrubbings and fussings, much less crude analogues based on evolutionary extinctions.

Thanks to these new techniques, researchers could identify episodes of climatic cooling and warming, which usefully matched the rhythms predicted by the Milankovitch pacemakers. They helped identify the onset of global cooling that preceded the first massive glacials. They corroborated evidence otherwise recorded by raised beaches, cool-water forams, and δ-O-18, which reflects the relative sequestering of oxygen isotopes in ice. Helpfully, a major boundary in the paleomagnetic chronology named the Gauss/Matuyama reversal became evident at the same time, and a similar chronicle in the oceanic crust allowed for correlation between the onset of the Pleistocene and the Olduvai Normal Event, a global geophysical marker. Still, the markers were time transgressive, as the jargon went, which meant that the originating events did not instantly leave a geologic signature; there were lags between the first measurable signatures of cooling and a full-blown glaciation. Still, it seemed possible at last to bring the Pleistocene into a geochronology based on absolute physical measurements.[17]

For 176 years after Charles Lyell had identified and named the Pliocene, and 170 after he coined the term Pleistocene to replace Newer Pliocene and Post-Pliocene, the marcher lords of geology had fought over where to draw the boundary between the two epochs, which is to say, how they might characterize each and their relationship. But from its outset the Pleistocene had resisted the demand that it be like all the other realms of geologic time. It was proudly, stubbornly anomalous. It demanded, as it were, a separate creation based on its own peculiar traits as it defined them. It was the epoch of ice and human origins.

When, in the early twenty-first century, the International Commission on Stratigraphy tried to abolish the Quaternary as an obsolete term for an idiosyncratic period, the Quaternary community rose in protest. Instead of disappearing, the Quaternary expanded in what some observers sourly called a "land grab," as the Pleistocene encroached into the Pliocene to the amount of 800,000 years, which put its official start at 2.6 million years ago and left the Pleistocene 260 times longer than the Holocene. The Pleistocene had proved as unstable as its climatic rhythms, spasms of extinction, and self-conscious hegemon of a species. In 2009, bowing to the Quaternarists, the International Commission on Stratigraphy approved the change by a vote.[18]

The poll fixed only one border. No less sticky was specifying that other, more recent boundary. According to the mechanisms of glaciation as mediated by Milankovitch cycles, the glacial epoch has not ended; it only paused about 12,000 years ago for an interglacial that, according to the mechanisms of the astronomical model, should even now be spiraling back into a new glaciation. By the logic behind its new origin, the Pleistocene should not have ended but continues unabated into the present day. The astronomical pacemakers granted no basis for the Holocene, certainly not as an epoch of equal stature with the Pleistocene. Logically, the Quaternary was either one epoch with internal divisions or two, divided according to common principles. The Quaternary, however, rather like its prime themes, seemed to obey another form of reasoning.

In practice, the borders were as jumbled as a terminal moraine. At the Pleistocene's upper boundary the paleontological record reasserted itself, as it pointed not to recurring rhythms but to a radical rupture and a departure. The rupture was a mass extinction, most spectacularly of larger mammals. The departure was the emergence of one species, *Homo sapiens*, that set about restructuring the planet and even perturbing the climate, with consequences as extensive and profound as the ice. These events destabilized the climatic chronometer and left the Pleistocene with one index for its opening and another for its closing. It was all arbitrary, and from time to time the solution was jeered at by other geologists, who threatened to drive the Quaternary System as a chronostratigraphic unit toward the same kind of oblivion recorded for its mammoths and woolly rhinos. They wished the Quaternary as a unit of geologic time might go the way of Pluto as a planetary body.[19]

Quaternary researchers retorted that ice and people were not only too big and unusual to overlook, but they offered an organizing principle of their own. If they contradicted themselves, they might reply after Walt Whitman, that they were vast and contained multitudes. The organization of past time could take many forms. The Pleistocene proposed another kind of logic than the precepts of the ICS-sanctioned Geologic Time Scale. It implicitly proposed narrative as an explanatory schema. It told a story.

Why should the names and dates so matter? Weary critics have dismissed the squabbling as mere semantics, quibbles over names; the rocks endure. What earth science needs ideally, they insist, is a global chronicle—a physically based planetary chronometer not dependent on the vagaries of eccentric and contingent processes of erosion, deposition, and evolution; a universal timescale not tied to particular type sites; and an absolute reckoning of history, not one defined by relative proportions. Geology should look to strict temporal meter sticks that would permit a complete record of Earth events constructed according to common principles.

Yet time *is* relative; it is as dense or thin as the events (or the evidence of events) make it; there is no absolute chronometer by which to record its ticking. Moreover, a chronicle is not a narrative. A chronicle is data, however complete and referenced to physical timepieces and fossil-correlated strata. The organization that matters is one that invests a timeline with significance to people, that endows it with meaning, that transforms chronicles into narratives that bestow on events a theme and an organizing principle that gives normative shape to an empirical tumult of bones, stones, air bubbles, and other hodge-podge from the past. Wallace Stegner once observed of landscapes that "no place, not even a wild place, is a place until it has had that human attention that at its highest reach we call poetry," until the "things that have happened in it are remembered in history, ballads, yarns, legends, or monuments."[20] That is not a bad description of what has happened to that wildest of past places, the Pleistocene, and why it can defy the strict logic of geochronology. It overflows with geologic monuments, lost tribes of creatures, and stories.

For narrative to shape such meaning requires well-defined beginnings and endings. Paradoxically, a reliance on "the science" may actually destabilize such a narrative. Build on the sands of data, and you will be swept away by the next flash flood of discovery. That is exactly what has delaminated so many creation stories in the modern era. It is not just that they fail to cope with scientific facts, but that the nominal facts so often change, or shape-shift in meaning as their context changes. Such is science's velocity that today's startling revelation quickly becomes yesterday's data point. A metacarpal in Siberia announces a new species. A humerus fragment in the Sahara repositions the development of bipedalism. Ironically, classic Greek myths show more staying power—are more likely to be read—than past scientific theories. Aesop's fables continue to speak truth about the world in ways that pre-Socratic science cannot, because they address enduring ethical questions in ageless ways. As a creation story, the Book of Genesis became a problem only when it was taken as a literal account of the natural world. But the same can be said of Louis Agassiz's cosmologies or those Darwin-inspired paleontologists such as Henry Fairfield Osborn, which now appear as dated as Herodotus's *Histories*. Hypotheses are as disposable as laboratory pipettes; today's positivist theories become tomorrow's folktales. To be powerful, a narrative must instead be anchored in art and philosophy, since aesthetic closure and moral resolution are what convey the context that endows facts with enduring meaning. They make a chronology into a story and distinguish a narrative from a mathematical model. The quarrel over the Pleistocene's borders is a contest over narrative, and it is not one that the keepers of the geological timescale can fix, with or without voting.

Pleistocene Palimpsest

The physical Pleistocene is a palimpsest, eroded and filled and emptied again—erased and overwritten—with each surge and collapse of glaciation. Likewise, its informing concepts compound, one atop the other. Which of them matters, and whether any can master the others, will depend on context and purpose. What grants intensity to the public discourse over obscure indices of oxygen isotopes, limestones in Calabria, and deep-ocean forams is the fervor of its implicit subtext: this is a story about us, not indirectly by describing the natural world in which we live but about our genetic origins and core identity as creatures of Earth. The problem with the Vrica horizon was finally not that it was too limited in the specifics of its locale, but that it was too constrained in the reach of its moral imagination.

Two great disturbances still give the Pleistocene its shape. Its Ice Age came with the rarity and impact of a colliding asteroid; and it is possible to date its origin, decode its master plot and subplots, and isolate the sweep of its story from among the quotidian of routine Earth events. So, too, the emergence of *Homo* injected a new perturbation into the chronicle of Earth time, one of equal, and perhaps greater, force. The species has ranged as widely as the glacial and periglacial landscapes, it has acquired a power as great as the Milankovitch-mediated insolation, and, singularly, it can contemplate and compare its own presence on the planet. Modern humans can, for example, invent an epoch of geologic time that defines them as they choose to be defined.

The most unsettling issue is not whether the story selected is arbitrary but that its protagonist is the arbiter of that story and lacks an outside referent. Compared to such perturbation, the stretches, tilts, and wobbles of planetary orbits can appear quaint, like relics from an age of absolutism when time was unchangeable and prime movers remained securely in the heavens. Yet this peculiarity is what has made the Pleistocene into something other—something more—than a slice out of the geologic time scale. It has made an epoch into an idea, and since that idea has changed over time as much as the material subject it describes, into a narrative.

Chapter 3

Story

The historian, like the physicist, lives in a material world. Yet what he finds at the very beginning of his research is not a world of physical objects but a symbolic universe—a world of symbols.

—Ernst Cassirer, *An Essay on Man* (1944)

FROM THE FIRST REVELATIONS—the encroaching ice of Alpine valleys, the excavation of monstrous bones—the Pleistocene was a long-ago time joined to the present by a story. A narrative organized around some informing principle (or insight) connected the sherds of the past. And while the possible stories of the Pleistocene are infinite, several have laid major claims, and two dominate. One, grounded in ice, is a geographic narrative. It looks outward and emphasizes, the environmental setting and how it acts as a stress or forcer to establish new species, *Homo* among them. It is a story in which climate turns the kaleidoscope of Earth elements into novel patterns. The other, anchored in the hominims, is a genomic narrative. It looks inward and accents genetic inheritance as the source of enduring traits and so traces the emergence of *Homo* out of the clade of primates. It is a story in which natural selection acts on a genetic inheritance constantly tweaked by new mutations.

Each of the dominant narratives has its own source fossils, its own version of evolutionary development, and its own larger vision of how nature operates and how hominins fit into the great scheme of things. Both make *Homo* a survivor and a striver, but they locate the source of its distinctiveness and the motive force behind its character differently. If pushed, proponents of the two perspectives would surely argue that they are one and that they differ only in matters of emphasis. In each narrative *Homo* is the outcome of both factors—genes and geography, paleoanthropology and environmental science—in dialectic. Each simply locates the narrative fulcrum in a different place.

Yet the differences between them are real. They reflect not only distinct research traditions with their own techniques, concepts, evidence, and history but contemporary customs and perhaps surprisingly (or perhaps not) a reincarnation of ancient traditions of thought. The geographic narrative relocates an unmoved mover from the heavens to the atmosphere. The genomic narrative finds a new locus for the defining spark within. And in expressing a vision of what makes us human, they move beyond excavated bones and the extraction of DNA into a moral universe for which the preferred medium is a narrative.

The transmutation of data into cultural meaning, however, is itself a fundamental feature of what we have come to call the Pleistocene, as powerful and strange as the jelling of ice sheets out of air. There are no embedded signposts in the geologic record that announce when the epoch begins and ends. Such identification is an act of cultural invention, a story of creation that has its own narrative, somewhere between a storied chronicle and a running gloss. It tells how ideas about geography and genomes begin, evolve, compete, and survive: how it is that modern humans understand the interaction of the geographic and genomic narratives and how they interpret the character of narrative itself. It tells the story of how the intellectual epochs that have come to scrutinize the Pleistocene, including modern science, have happened. Unlike the others this narrative must reflect consciously upon itself.

The Geographic Narrative: From Big Chill to Spark Within

When the Pleistocene opened, its geographic and genomic narratives met most powerfully in Africa. Here, according to contemporary understanding, is where the strike of climate on primates cast the spark that led to humanity.

Such a conception, however, would sound odd to the conceptual framers of the Pleistocene. Africa had little ice, and it seemed remote from the sources of human origins. Almost alone, Charles Darwin, pondering biogeographic considerations, argued for Africa as the hominin hearth. "In each great region of the world," he observed, "the living mammals are closely related to the extinct species of the same region. It is therefore probable that Africa was formerly inhabited by extinct apes closely allied to the gorilla and chimpanzee; and as these two species are now man's nearest allies, it is somewhat more probable that our early progenitors lived on the African Continent than elsewhere."[1] Most scholars looked not to primate relatives but to cognate civilizations. They imagined Africa as an outlier, not a place of origin.[2]

African Pleistocene

Among the continents today Africa is the highest, the warmest, the most tropical, and perhaps the most prone to seasonal wetting and drying, and the one that displays the most powerful mixing of the symmetrical with the unbalanced. The equator divides it nearly equally north and south, but the north is large—a bulge of crust the size of Australia—while the south tapers to a blunt point. Between the two regions, resembling a double hinge, are exaggerated highlands and lowlands, the Ethiopian plateau and the Congo Basin. A similar pattern applies to climates and biomes—symmetrical north and south and asymmetrical east and west. There are mediterranean climates at the far south and far north; then come deserts, grasslands, woodlands, approaching north and south to a common center of rain forest. The real asymmetry is between the lands east and west of the Rift Valley. The east belongs with the monsoonal climate of the Indian Ocean; the west, with general circulation and the seasonal migration of the intertropical convergence zone. The Rift tears across these otherwise orderly progressions.[3]

During the Pleistocene Africa was much different environmentally. There was little direct glaciation. The highest mountains—Kilimanjaro, Mt. Kenya, Ruwenzoi Aberdares, and Mt. Elgon—amassed more than 300 square miles of ice, while the Ethiopian plateau had 290 square miles of ice. And there is some possibility of glacial ice in the Saharan

Tibesti and the Cape mountains. As elsewhere, the ice mounded and melted repeatedly. The real impact came indirectly from the glaciation of Europe and the polar seas. The Eurasian ice dome deflected the circulation of the jet stream southward and sharpened the temperature gradient with the tropics, while the icing of the oceans did something similar for ocean circulation, and more chilly waters along the African coasts reduced evaporation. The rending Rift and Asia's rising Tibetan Plateau strengthened these trends. The upshot was a great drying of the continent.[4]

Instead of ice sheets, Africa had pluvial lakes, which filled and drained with the Pleistocene's climatic cycles. Some of the lakes were immense: Lake Chad and Lake Palaeo-Makgadikgadi together totaled 150,000 square miles, and there were many other, smaller bodies that swelled out during the wet periods. The pulsing of flooding and drying left lung- and mud-fish as the only endemics in the Nile and Great Lakes and marooned crocodiles in Saharan oases. And instead of ice lobes and loess plains, Africa had deserts—great ergs, dune fields, and sand velts. These, too, swelled and shrank like a vast biogeographical bellows. Some so crowded grasslands and rain forests that those biotas shriveled into narrow belts along the littoral of the Gulf of Guinea or retreated into protected refugia; others overran woodlands, pushed into what today are rain forests in the Congo, and sent streamers of sand to block the Niger and the Nile.[5]

While the details are unclear, and may never be known, the upshot is that the rhythms of the African Pleistocene expressed themselves largely in long wet and dry periods of which today's seasonal cadences are a diminutive echo. As the northern continents flipped into and out of ice, Africa switched into and out of aridity.[6]

Ice, water, sand—all can come and go under strictly mechanical formulas. One cycle can replicate another. But warming does not act on a biota as sunlight does on a block of ice, melting a solid mass of rain forest into flowing pools of savanna. It has to be transmitted through the complex channels by which climate becomes weather, and it has to pass through ecosystems in which the grazers, browsers, and decomposers, and the predators that help control their numbers, all dampen or elevate the capacity of the biota to respond. There is little reason to believe that each revolution of chilling and warming only recapitulated the old communities. Rather, their repeated climatic firing and quenching left them leached, culled, bulked up, and toughened.

The dominant story—the evolutionary epic, as it were—opens with the Pleistocene's climatic inflection to cold and dry and its biotic movement from forest to grassland. The geologic record in Africa documents a shift toward both some 2.5 million years ago: savannas become more prominent, crystalizing out of rain forest and woodland, and a fossil record parallels that transition, from forest to grassland species of bovids (antelopes, specifically). The transformation stretched from Ethiopia to South Africa, tracking along the Rift Valley and perhaps catalyzed by it. In 1985 Elisabeth Vrba formally expressed this observation as the "Turnover Pulse" hypothesis, which argued that climate catalyzed a massive turnover of the landscape and led to a comparable turnover of species. The event was large and critical, affecting most of southern Africa.[7]

The most interesting point of contact between the Pleistocene's two informing narratives is that antelopes comprise 60 percent to 80 percent of all mammalian fossils in the relevant strata, and thus serve as an index for climate and evolution. What happened to bovids thus also happened to hominins. They had to adapt to grasslands, or to learn to move back and forth from forest to savanna, or to find ways to exploit the jumbled assemblages of woods, grasses, and shrubs that, like the Rift, rumpled the bio-landscape. It seems that precisely at this inflection point, a "center of endemism for the hominids" (or at least for the preservation of hominin fossils), the first example of *Homo* appears.[8] Darwin's bold intuition, based on biogeographic factors, that Africa was the hominin hearth has proved correct. The selective power of the geographic narrative is one reason why.

Turnover Pulses: Changing Climates of Opinion

Although labeled a hypothesis, the Turnover Pulse concept acquired the power of a master argument by presenting the "habitat theory" as "an alternative to the competition paradigm." In announcing it, Elisabeth Vrba explicitly presented the idea as something beyond the mundane workings of normal science, and accordingly reached outside the usual realm of scientific authorities for support. In particular, she appealed to the eminent philosopher of science Karl Popper to assert that "bold ideas, unjustified anticipations, and speculative thought, are our only means for interpreting nature," and she offered historical examples of just such bold assertions. Big change required some force majeure applied from outside the discipline: that was as true within science as within the natural world that science described.[9]

Implicit in this appeal is a correspondence between how nature evolves and how science evolves. Popper brushed aside the prevailing notion that science advanced "by gathering new perceptual experiences, and by better organizing those which are available already," not because it was wrong but because it "seems to miss the point." Nature does not give answers: it must be pressed in order to yield. Experiments are struggles. Understanding is ephemeral. We cannot know irrefutably, only guess, and "our guesses are guided by the unscientific, the metaphysical (though biologically explicable) faith in laws, in regularities which we can uncover—discover." Absolute certainty, paradoxically, can come only from "our subjective experiences of conviction."[10]

Yet this is what the Turnover Pulse proposed for nature. It argued that an exogenous stress, a major shift in climate, had impressed itself on southern Africa and compelled the biota to change. It had shaken and pried apart the old patterns and sifted the pieces into new arrangements. So, too, did science seem to operate. It moved hugely only when pressured from the outside, from a change in the climate of opinion. The accretion of data, like a drizzle of mutations in the gene pool, could not produce breakthroughs. Ideas of nature, ideas of science—each seemed to inform the other.

Popper's *The Logic of Scientific Discovery* was translated into English in 1959 with a new edition in 1968. Between the two editions Thomas Kuhn, a noted historian of science, published *The Structure of Scientific Revolutions* in 1962, and in 1970 released an expanded second edition. The motivating prompt behind Kuhn's study was the recognition that the "concept of development-by-accumulation" that characterized most histories of science did not explain the spasms—the upheavals in understanding—that characterized most narratives. Kuhn thus distinguished between a "normal science" that plodded through a routine agenda of puzzle solving and "revolutions" that leaped into new realms, established new "paradigms" for research, and spawned novel species of disciplines. The catalyst for change was an "anomaly" in the form of data and ideas that didn't seem to fit and that stressed the landscape of thought and forced a rapid reconstitution.[11]

The book became a sensation. It marked itself a paradigm shift not only in thinking about science but in understanding how the world worked generally. With the turbulent 1960s as a cultural backdrop, Kuhn's depiction of anomalies and revolutions became, for many intellectuals looking for radical reform in their fields, both an anatomy of revolution and a prospectus for it. It helped validate a vision of a world in which long periods of stasis might shake free into sudden spasms of upheaval or, to rephrase that observation, in which eras of evolutionary somnolence might rapidly morph into pulses of overturning biotas. An older vision of historical movement—of gradual, cumulative, pedestrian evolution, of Darwin's claim that nature does not make jumps—seemed itself swept aside by visions of cataclysmic change. The simultaneous appearance of a chilling Earth, African rain forests dissolving into savannas, and the eruptive emergence of new species seemed to match with the novel perspectives of scholars that the macrocosm of natural evolution and the microcosm of human understanding could mutually align.

Few scientists avoided the shadow of the Kuhnian paradigm. Its influence fell especially on those eager for intellectual turnovers and those who sought alternatives, and even on those who proposed self-proclaimed hybrids, such as the concept of punctuated equilibria. The latter's authors, Stephen Jay Gould and Niles Eldredge, asserted, with a boldness that matched the era's fascination with disruptive social movements, that punctuated equilibria "dominates the history of life."

The long narrative of organic evolution on Earth experienced most of its defining changes in "very rapid events of speciation." Moreover, they made explicit that their perspective applied to human evolution as well as to horses and *Hyopsodus*. "The record of human evolution seems to provide a particularly good example: no gradualism has been detected within any hominid taxon, and many are long-ranging." The data from bones matched that of molecular genetics. Nature leaped, and *Homo* sprang from the end of one such jump. The Turnover Pulse grounded that assertion in a time and place.[12]

Such assumptions—the organizing concepts, often metaphors, that arranged points of data the way constellations drew pictures amid the stars of the night sky—were in the air. But one leap can lead to another, or to a moment of rebalancing. Counter-theses immediately appeared, such as the Red Queen model proposed by Leigh Van Valen in 1973, which was also announced as a "new evolutionary law." This particular formulation placed the burden of evolutionary change back in the biota. What is most revealing perhaps is that it seemed possible, perhaps necessary, to counter bold new ideas about bold rapid change with other bold new ideas. Whether or not nature moved in pulses or according to revolutionary manifestos, ideas about it certainly did.[13]

In the end, what matters is less the specific argument, or whether evolutionary change seeped or leaped, but how and why the explanation is believable. Such ideas have to speak beyond their subdisciplines: society supports those enterprises precisely because they promise to align with what William James called the felt world. They speak in the language, metaphors, tropes, and experiential sense of their larger culture. It is "particularly in periods of acknowledged crisis," Thomas Kuhn observed, "that scientists have turned to philosophical analysis as a device for unlocking the riddles of their field. Scientists have not generally needed or wanted to be philosophers." At issue is not simply whether the science is popularized but whether it makes sense even to its own practitioners. Nature only replies to the questions we ask, Popper noted. The questions we choose to ask come from our larger sense of how we understand the relationship between us and nature. In Popper's phrasing, "[W]e have to 'make' our experiences." And we have to endow them with meaning: "[O]nly in our subjective experiences of conviction, in our subjective faith, can we be 'absolutely certain.'"[14]

What experience and faith averred was the sense that geography and genomes had interacted around 2.6 million years ago to create a genuine horizon in geologic time that might also serve as its own metaphor of change. An epoch later humanity could look back and construct a story of how it had all happened.

The Genomic Narrative: Not in the Stars, but in Ourselves

The genomic narrative has evolved in parallel, almost as an intellectual homology, like the convergent development of wings on bats and birds. In this conception we are what our genome, the biochemical repository of the genes we have inherited, makes us. The central dogma of molecular genetics is that individual genes code for specific proteins through mechanisms of command and control. The environment can only pare and push what our genetic heritage makes possible. Know its genes, and you know the essence of a species. "The fault," as Cassius informed Brutus, "is not in our stars, / But in ourselves."[15]

Principles and Primates

In this perspective, what distinguishes *Homo* from its closest creatures is best measured not by shape or behavior but by the differences in their genomes. Paleoanthropologist Richard Klein has observed, "To the extent that fossil forms resemble living ones (or each other), it is often difficult to determine if the shared features are actually primitive characters inherited from a relatively distant ancestor or, alternatively, are parallelisms that genetically similar taxa developed independently as they adapted to similar circumstances."[16] More simply, modern humans are not closest to those creatures with whom they share similar behavior or occupy common ecological niches but to those with whom they share the most genetic material. The ancient Delphic challenge to "Know thyself" was an injunction to look inward to find the essence of character. Today that oracle would likely be restated to read "Decode your genome."

A genome records its own past history; its molecules are an archive of evolution. This realization, combined with techniques for extracting such evidence, grew in the 1960s. It began by screening and comparing blood-serum proteins among the higher primates, and at the hands of Morris Goodman, the proteins were arranged into evolutionary trees. Out of such experiments arose the realization that the rate of change in serum albumins showed a temporal regularity, as the rate of neutral mutations in these proteins established a kind of biomolecular clock. In a series of papers published between 1966 and 1967 Vincent Sarich and Allan Wilson expanded these biochemical findings into a molecular phylogeny that allowed researchers to construct a hominin genealogy. As technology further improved, researchers went from analyzing proteins to analyzing the informing genes themselves. The greater the difference in their DNAs, the longer two species have existed apart. It was then possible to compare the genomic map of hominins to their geographic map. The two narratives twisted together, not unlike strands of DNA.[17]

Because it addressed history, the genomic narrative continued to appeal to paleontology in addition to biology for its descriptive language. The molecules were a novel kind of "fossil." They created a new variant of chronometer and they expressed themselves in the records of evolution. Yet the dates discerned by those who studied the morphology of fossil bones and those who studied the fossil molecules coded by DNA did not agree. The shapes of bones, which had always seemed the most immutable of data, became diagnostically unstable; form might not always indicate function, and in particular it might not indicate location in an evolutionary tree. In 1971 Sarich proclaimed that "one no longer has the option of considering a fossil specimen older than about eight million years a hominid *no matter what it looks like.*" Then in 1987 it became possible to add mitochondrial DNA to the molecular model. The debate moved from the origins of the earliest *Homo* to the appearance of the latest, *Homo sapiens*.[18]

There were plenty of clashes between the various approaches, and numerous occasions when dates did not agree. Some followed a familiar formula, as each research group clung to its own evidence, analytical techniques, and styles of explanation. But more was at issue than disciplinary methodologies. Each group necessarily defined humanity differently. If the genomic chronology did not match the geographic one, it was because their definitions of what it meant to be human were out of sync. The choice of techniques brought with it different options for defining *Homo,* and such choices determined the narrative that could result.

The payoff was not in DNA sequences but in the behavior such changes implied, because this was the real index of a moral world. The premise is that we differ from other creatures not because we look different but because we act differently, and we act out of learning rather than instinct alone. In this regard the genomic narrative was only a modernized version of a very old discourse, in which we distinguish ourselves from the beasts by our reason, our capacity to choose and act freely, and our possession of an intangible something that other creatures lacked. There was some spark within that separated us from the rest of creation. While there was no code in the genome for soul, the need for it remained, it seems. If it didn't exist, it would have to be invented.

Man the Hunter: Founding Flaws

The genomic narrative also had to adapt to its cultural setting. Unlike the geographic narrative, its logic required that it locate the motive agency within, and it found an illustration in the evolutionary record of prey and predators. In the South African cave Swartkrans, the lower strata show australopithecine hominins as prey; they mix with the bones of baboons and antelopes. In the upper strata that relationship inverts, and early *Homo* controls what gets deposited. Something had changed; some behavior had allowed prey to become predator to the hunted to become the hunter. The genome now seemingly coded for new, defining traits.

For a while, as paleoanthropologist Matt Cartmill has elaborated, the hunt became an informing principle in the search for a behavioral separation of humans from the rest of nature. So deeply embedded was the evidence for it in hominin history that it seemed hunting could only be explained by appeals to genetics. What made hunting especially diagnostic was that humans could turn it on themselves in the form of war. Hunting—hunting as humans did it—was an index both of humanity's unique talents and of its inherent flaws. "Whether we love hunting or hate it, eulogize its blinding passion or condemn it," biologist Valerius Geist asserted, "hunting was the force that shaped our bodies, molded our souls, and honed our minds." The movement reached a high-water mark in 1966 with a major symposium on the theme of "Man the Hunter," which followed a previous conference on the "Origin of Man."[19]

In the summary essays there was little doubt about the assumed significance of hunting. It was, W. S. Laughlin declared, "the master behavior pattern of the human species." It was the "organizing activity" that integrated anatomy, physiology, genes, and mind, and it spanned the "entire biobehavioral continuum" of humanity from individual to community. Sherwood Washburn and C. S. Lancaster went further, if that is possible, and argued that "to assert the biological unity of mankind is to affirm the importance of the hunting way of life. . . . The biology, psychology, and customs that separate us from the apes—all these we owe to the hunters of time past." Further, hunting defined the "whole human view of what is normal and natural in the relation of man to animals"—and, one might add, to one another, since "until recently war was viewed in much the same way as hunting."[20]

So fundamental a trait had, it seemed, to be hardwired into human biology. Conveniently, the genomic model was firing its first salvos as the symposium proceedings got into print. But the two enterprises shared an ancient tradition that identified both the success and the fatal flaw of humanity in the nature of its separation from the rest of creation. That convention had altered over the centuries, but the belief that hunting *qua* warring was a kind of original sin seemed plausible to a civilization that had ripped itself apart with two world wars and was now committed to an apparently endless series of conflicts within a cold war, of which the Vietnam War was the most recent avatar. From the moment of its genetic creation, humanity had been intrinsically violent—a hunter, a warrior, a killer. "Since 1960," notes Cartmill, "the picture of *Homo sapiens* as a disease of nature—a mentally unbalanced predator threatening an otherwise harmonious natural order—has become so pervasive that we scarcely notice it anymore." On this, at least, paleoanthropology and its larger culture agreed.[21]

Over time, a revulsion set in, based not merely on new evidence but on new social norms and perspectives that did not wish to have humanity burdened by such an unsavory and seemingly immutable heritage. American society spawned a counterculture based on a non-militaristic ethos; Europe buried its ancient blood feuds to forge a political union based on consensus and collective values; women became a major political and economic presence. Instead of hunters, we were gatherers; instead of fighters, sharers; instead of toolmakers of weapons, artisans of jewelry and musical instruments; instead of selfish cunning, a social cleverness. This view, too, will no doubt shift as discovered genes and invented culture recombine in the future.

What endured is the query itself, and the genres by which it is expressed. The characters change, the plot pivots differently, the moral setting morphs, but the storytelling goes on.

The Cultural Narrative: The Gloss of Temperament

In discussing why some ideas prove acceptable to some people and not to others, William James noted that the "history of philosophy is to a great extent that of a certain clash of human temperaments." At issue is not a particular hypothesis and its supporting logic and evidence but a cast of mind. "The one thing that has *counted*" is "that a man should see things, see them straight in his own peculiar way, and be dissatisfied with any opposite way of seeing them." This predisposed temperament is what "really gives" a stronger conviction than any of the "more strictly objective premises" advanced.[22]

"He *trusts* his temperament," James continued. "Wanting a universe that suits it, he believes in any representation of the universe that does suit it. He feels men of opposite temper to be out of key with the world's character, and in his heart considers them incompetent and 'not in it,' in the philosophic business, even though they may far excel him in dialectical ability." What James describes for an individual can apply equally to whole clades of researchers, whose felt sense of the world will change from time to time. James argues that although the temperament—the temperament of the times—carries the greatest conviction, for it is "the potentest of all our premises," it is never mentioned, because it can make no public claim to "superior discernment or authority." The result is that public discourse displays "a certain insincerity."[23]

Few Pleistocene researchers would accept such a formulation. They will insist that as scientists they are exempt from such notions, that the discipline of science will strip away such hints of subjectivity, that science has triumphed precisely because it has sheared away such woolly sensibilities as philosophy, art, and temperament. T. H. Huxley once argued that science would "sooner or later" become "possessed of the law of evolution of organic forms—of the unvarying order of that great chain of causes and effects of which all organic forms, ancient and modern, are the link." So today's scientists would likely believe, if not to proclaim so blatantly, that they will fill out, missing link by missing link, the great chain of knowledge. Disagreements reflect only undiscovered data, sibling rivalries, and misconstrued and simplistic ("legendary") conceptions of science. Eventually, as Huxley and his positivistic progeny insist, what had been myth and philosophy will become science or be dismissed. What had been stated in narrative will be reforged into the theories, models, and even the laws of exact science.[24]

Yet is it doubtful the public supports such research so enthusiastically because it wants to know whether a particular bone was scratched by a hyena or a stone flake, or care how specialists distinguish processual from postprocessual archaeology, like taxonomic splitters multiplying breeds of brown bears, or whether they are a genuinely new species of thought. They want to know who we are, how we came to be, and how we might act now and in the future. These concerns are, in fact, embedded in the nominally scientific tropes that dominate Pleistocene studies. The geographic narrative presents early ancestors as marginal, even hapless, creatures ever vulnerable to the whims of climate and geology, yet creatures who could transcend the challenges of their environment through luck and pluck. The genomic narrative imagines powerful but imperfect creatures whose instincts could drag them to ruin but whose divine spark, however secularized, pulled them toward something like salvation; it tends to argue for the difficulties of overcoming inherited traits.

Such matters are, at heart, an argument about behavior that bones and genes cannot, unaided, answer. They have more in common with Hamlet pondering Yorick's skull than the lab tests of zooarchaeology. Many practitioners of Pleistocene studies would like to see such spongy soliloquies go away. (So would practitioners of other fields as well, but Pleistocene science can strip the tendency out only by ignoring humans, which would remove a defining property of the field.) They would prefer to treat hominins with the same nominal objectivity as they do equids, elephantids, and ursids—without the refracting prisms of culture. Such creatures may be turned into symbols and fables (and have been), but a natural science is possible for them in ways that appear impossible for humans. The history that most interests humans is precisely where matter and mind fuse.

To those intent on crafting the discipline into a stronger science, the lingering role of the humanities is an embarrassment and an encumbrance, or at least, a worthless vestige, like an appendix that ought to be removed before it becomes a point of lethal infection. They doubt narrative itself—would like to dismiss not just this narrative or that one but the very legitimacy of narrative as a mode of explanation. They know that all narrative is not equally useful, but they forget that narrative, too, has its scholarship.

Narrative Arche: Leap of Imagination

Since ancient times thinkers have sought not only to use narrative to express ideas but to understand how narrative itself works. The study of narrative is as old as the earliest recorded natural philosophies. In his *Poetics* Aristotle focused on the concept of *muthos,* which fuses plot and fable. Plot gives figuration, or a particular ordering in time, to events, characters, and places. Fable endows the story with moral meaning, in that it instructs, in the process of unfolding revelations, about universal aspects of the human condition. It makes us care about the outcome. Depending on its particular genre, it allows laughter, catharsis, knowledge.

The quest for narrative and the quarrel over it thus date back to the origins of Western thought. In the sixth century BCE, Thales of Miletus conceived that the world had arisen from the action of an organizing principle, what he called *arche,* on a foundational substance, which he imagined to be based on a watery medium. Arche connoted both origin and order, for it was an informing principle or organizing conceit applied to nature, and as such it lent itself readily to transcription into narrative. The Ionian natural philosophers who followed him subsequently identified various means that acted on assorted matter, but the process remained the same. They distinguished themselves from theologians in that no spiritual presence served as a causal agent. They shared with the cosmogonists like Hesiod the conviction that the outcome was a kind of story whose beginning was determined by its conclusion.

Their success in paring the spiritual—broadly conceived—from the material made their schemas an early science. But that success also divorced their inquiry from ultimate questions and moral scholarship. It was precisely this divorce, in the conception of F. M. Cornford, that turned Socrates to a philosophy that did not concern itself with natural causes and effects but with questions of what makes a good life. In Cornford's scenario science did not grow out of the failures of early philosophy; philosophy grew out of the failure of early science. It had to address those vital issues that science could not or would not.[25]

Creation stories themselves began long before modern science, and much of what modern science, certainly paleoanthropology, has done is to shake out their old contents and replace them with new pieces while keeping their form. The narrative form itself conveys much of the intended message. The shape of these origin narratives makes them myths, which is to say, a story that conveys themes and values of significance to its society. Stuffing a cosmogony with data rather than dreams may make it science-informed; it does not make it science. It belongs with philosophy and literature, and that is why it is both interesting and necessary.[26]

A narrative understanding, as Paul Ricoeur has observed, is "much closer to the practical wisdom of moral judgment than to science, or, more generally, to the theoretical use of reason." But narrative has an even more fundamental role. Self-understanding finds expression in a "narrative identity": "[W]e learn to become the *narrator* and the hero *of our own story*, without actually becoming the *author of our own life*." The study of human origins extends this understanding to the life of humanity overall. It describes perfectly the way the epic of human evolution, of how we have become what we are, is presented whenever scholars attempt to move beyond the rendering of lithics, sherds, mandibles, and climate recordings on ice and rock. It is why the culture cares about the quest.[27]

A proper narrative requires closure of two kinds. It needs explanatory closure, in that it must account for what happens and why, and it needs aesthetic closure, in that it must contain the relevant information and causes within its narrative arc. It must appear to be both consistent and complete within the bounds of what it describes; it must satisfy as well as explain. Its arche is what distinguishes narrative from data sets, listed artifacts, chronicles, and annals. It's what allows the story to conclude with a sense of satisfaction that its tensions, conflicts, and moral meanings are resolved.

What Pleistocene stories require, in particular Pleistocene creation stories, is a narrative arc that ends with humanity. It is our sense of who we are that will identify the telling traits, select the proper arche, and determine the historical span from the past point of origin to the present human condition. Yet narrative, no less than evolution, has its logic, and as the pre-Socratics realized, both basic substance and founding form must stiffen around some invariant property. The substance cannot morph endlessly, or it will assume no usable shape. The form cannot slide from one cause to another ceaselessly, or the arc (of causation and plot) can have no beginning or end. Ideally, somewhere, there must be an unmoved mover who acts on immutable matter.

In more modern times both the moving and the matter moved appear more complex and malleable, and less drawn from the realm of everyday experience. The substance may be dark matter and quarks rather than water or air, and the arche may be gravity or string harmonics. For human origins they may be genes and natural selection. Since the late Renaissance, however, modern science has been determined to pare away the question of ultimate causes and concentrate on proximate (or "efficient") causes. It deliberately repudiated the teleology that underwrote Aristotelian natural philosophy and Judeo-Christian theology. It sought to explain how particular phenomena worked, not why creation happened or where it might trend. It segregated cause from purpose, the morality of action from its mechanics, and the workings of nature in isolation from its ultimate unity and meaning. For the naturalist, the outcome is not foreordained and history does not advance toward a predetermined end.

But the culture also wants the kind of meaning that story brings, and for this to happen there must be limits to the plasticity of stuff and its organizing conceit, bounds of an alpha and an omega, or the narrative won't work. For the storyteller, design must be present. That's what makes it art and not science. In brief, it is less the understanding of our origins that explains our identity than our felt identity that shapes the understanding of our origins.

Today, there are as many competing explanations for humanity's beginning as a species as there are bustling, jostling hominins to explain. Like the cosmogonies of the pre-Socratics, each accents a peculiar feature, be it tools, bipedalism, a big brain, or a distinctive behavior. The geographic narrative emphasizes our ability to adapt and change and points implicitly into the future. The genomic narrative underscores our capacity to endure, and the limits imposed by our past. All are modernized versions of the arche as they seek out originating causes and an organizing form by which to express it. They are, in a sense, the archetypal sciences of the Pleistocene. And from the beginning they would be joined by a new field disipline. The same year Agassiz announced the Eiszeit, the term "archaeology" first appeared.

PART TWO

THE GREAT GAME

Let me begin by reminding you of the fact that the possession of true thoughts means everywhere the possession of invaluable instruments of action.

—William James, *Pragmatism* (1907)

Here begins the Great Game.

—Rudyard Kipling, *Kim* (1901)

The Pleistocene is a compelling epoch in which to observe the unfurling verses of what Rudyard Kipling in another context once called "the Great Game." From the vantage point of early twentieth-century India, the Game was the geopolitical contest played out between Russia and Britain. From the perspective of early twenty-first-century Africa, the Game could aptly refer to the far older geobiological contest among species as the Pleistocene emptied and opened lands for species to claim and colonize—and to the various competing explanations of those events, for the Great Game is played out among species of primates, equids, ursids, and elephantids.

The Pleistocene, as a moment in natural history, engendered, killed off, and intermingled species. They came, they competed, they climaxed, or they fell to the side. The entire tribe of hominins—the habilines, the erectines, the Neanderthals, the floresientines, the sapients, and the others—lost to the sediments of time. The elephantids—the mammoths, the mastodons, the African *Loxodonta* and the Asian *Elephas*. The equids. The ursids. Untold biotas assembled, dispersed, and vanished, and then regrouped.

The Pleistocene as an invention of cultural history went through the same process with ideas. As climates of opinion changed, as centers of research emerged and faded, the ancestral stock of concepts about nature and how it works, and about what makes humans what they are, underwent an intellectual evolution. Whole taxa of ideas and methods of inquiry came and went amid cultural upheavals. Ideal Forms. Chains of Being. Evolution by design and chance. Sciences like the newly invented geology and reborn biology; philosophies of positivism and pragmatism; the romantic and the modern. The discovery of Pleistocene extinctions, for example, caused inherited ideas about the permanence and stability of creation to disappear themselves. The vanished mammoth stood not simply as a species that had become extinct but as a symbol for the concept of extinction. The evolution of the horse became an exemplar for evolution itself.

For humanity the Great Game commenced when, as the Pleistocene dawned, the earliest specimens of *Homo* appeared and joined the scramble for Earth. The narrative of their ascension continues the contest through clades of competing stories.

Chapter 4

Footnotes to Plato

The safest general characterization of the European philosophical tradition is that it consists of a series of footnotes to Plato.

—Alfred North Whitehead, *Process and Reality* (1929)

FEW FIELDS can claim disciplinary dynasties equivalent to the Leakey
family in paleoanthropology. Louis and Mary Leakey's work in East
Africa, particularly at Olduvai Gorge in Tanzania, created a school of
research in and of itself. Louis's passion for fossils and the grandiose,
combined with Mary's skill and precision as an illustrator and archaeologist, made a forceful backdrop for the emerging field of paleoanthropology. Their children, Jonathan, Richard, and Philip Leakey,
have all, at some point in their lives, been involved in the processes of
fossil finding and interpretation. Indeed, some of the interactions
between Richard Leakey and other scientists are the stuff of modern
legends. Even today, two Leakeys, Meave (Richard's wife) and Louise
(their daughter), continue to do research in East Africa.

When Louis and Mary Leakey began their East African research
they were an enigma within the paleoanthropological world. Their
research foray into the Rift Valley was a first. Previous undertakings
had focused efforts in South Africa, China, or Southeast Asia. Louis,
however, maintained that his efforts and archaeological data would
"prove" Darwin correct in finding evidence for humanity's early evolution in East Africa. They began their efforts in Olduvai in 1951,
excavating tools in several beds of tuff. Finally their efforts yielded
a hominin, labeled OH5, but termed "Dear Boy" or "Zinj." (The

specimen was later described as a robust australopithecine from the Plio-Pleistocene border.) Their long isolation, showmanship, and startling discoveries have made them targets to vilify or mock, but they made fundamental finds and kindled public interest in ways that confirmed as them as a founding family of paleoanthropology.[1]

Their most celebrated find occurred at Olduvai Gorge, Tanzania, between 1962 and 1964. Appropriately it was a family affair: Jonathan Leakey found the bones; Mary Leakey oversaw their excavation; and Louis Leakey named and interpreted them. Their accepted ages were 1.9 million to 1.8 million years old, with a suggested origin for the incipient species around 2.4 million years ago. The fossil had some features of *Australopithecus* and some of *Homo*. The hominin was short, had long arms, a less prognathic face than apes, and a less capacious cranium than modern humans (630 cubic centimeters—roughly half). The most stunning discovery was the association, for the first time in the fossil record, of deliberately shaped stones beside the bones. Probably with advice from Raymond Dart, Leakey chose the name *Homo habilis*— meaning "skillful," or "handyman," or, colloquially, a "user of tools."[2]

True to Type

While the fossil was not the first evidence of very early hominins to be discovered, it posed the question of what that earliest human (*Homo*) might be. It marked a putative point of origin, a source for what would follow. But the handy habiline was still more useful. It established not only characteristics for the founding *Homo* but for the field that sought to examine him. The fossil became a type specimen and the story of its finding, a type narrative for paleoanthropology. Along with Raymond Dart's identification of the Taung child and Donald Johanson's later find of the australopithecine Lucy, it established the basic tropes of discovery. Even more, it demonstrated the role of types, or templates, in the process of understanding. What type specimens do for taxonomy, formal analogues do for comparative study and formulas do for narrative.

Traits and Types

This was the first hominin to be named *Homo*—a determination that has survived fifty years of scientific scrutiny. Why did it merit that classification? What traits did this creature have that not only differentiated it from earlier hominin genera like *Australopithecus* and *Paranthropus* but bonded it to us, as *Homo*? In other words, why was it more like us than like them and therefore earned the tag *Homo*? An answer will be fuzzy, as it must be as the source bones are old, fragmentary, and often deformed, and there are so few of them. They scatter on a phylogenetic tree like data points on a graph; they do not neatly align along a single curve; as data they connect only through leaps of logic, analogy, statistics, and faith.

But it is fuzzy, too, because the real issue is that people can only identify an ancestral *Homo* if they know what traits best or uniquely characterize contemporary *Homo*. The earliest explanations, by the field's founders, pointed to a suite of criteria including terrestriality, bipedalism, encephalization, and civilization, or more simply, to the recognition that early *Homo* lived more on the ground than in trees, walked upright, had a big brain, and relied on a social order informed by institutions, language, tools, and a moral code. The sequencing mattered, since how the traits unfolded would distinguish early version *Homo* from more recent models. Lacking much empirical evidence, however, authorities disagreed among themselves. Some, like Charles Darwin, thought the order of events would go from the trees to the ground, to increasing bipedalism, to a larger brain, to the tokens of civilization. Others, like Elliott Smith and Arthur Keith, thought it had to begin with a big brain, reflecting a Descartian perspective that reason is what most defines humans, and the application of reason, however primitive, is what surely propelled humanity's ancestors along the route of progress. Still others mixed the elements to tell the story differently, like children's alphabet blocks stacked and lined up in varied ways.

Further complicating the task, the morphological changes between species are incremental, a species has plenty of variation of its own, and the index traits do not evolve, each to its full extent, one after another. Bipedalism, for example, did not reach some maximum and then hominins turned to their next critical trait, improved that, and then moved to the next. Instead, all the characteristics continued to evolve in fits and starts, each affecting the others. Early *Homo* continued to acquire bipedalism as it grew a bigger brain, and as the cranium enlarged, the birth canal had to accommodate it, which affected the pelvis, which affected upright walking. Compared to a hypothetical progression, some traits could even appear to regress. The parts of the genomic mosaic would coevolve at varied rates.

Eventually more habiline specimens appeared from other African sites. They have tended to diverge more than converge, leaving uncertain just where the species borders might lie. Overall, they have shown a variability that is, in Richard Klein's word, "extraordinary."[3]

To the mind of Louis Leakey what most distinguished the early *Homo* sites was not the morphology revealed by the bones but the association of those bones with stone tools. This was the very basis for his taxonomic designation. Whatever swirl of properties was at work, whatever infinitesimal alterations in anatomy were manifest, the species had crossed a threshold when it began taking the raw stuff of nature and transforming it, not as most creatures did into a nest or niche, but into an extension of tooth and talon. What distinguished early humans was not their anatomy but their skills; they created technology. That was the defining, material difference between them and other hominins who could walk upright and live on savannas and grasp objects and had brains larger than chimps but who could not coordinate brain and hand to craft implements with which to extend their reach and compensate for weaknesses relative to other competitors. *Homo habilis* could.[4]

But with further discoveries so, it seemed, could other species—*Australopithecus garhi*, for example, which could push toolmaking back to 2.5 million to 2.6 million years ago, or further; and if one australopithecine could knap stones, perhaps others did as well. What the narrative asked for was a sharp point of origin, like a single spark that kindles a fire; what the sources gave was a scene of smoldering embers. Evidence of behavior, however, might substitute for evidence from anatomy. The implication is that we are not what we look like but what we do. Similarly, a narrative of discovery about the Pleistocene might substitute for a narrative of the Pleistocene itself. You could write a story of how people came to understand the field with more surety than you could tell a story based on the data from the field.[5]

The tools were not much. Some differed little more from naturally chipped rocks than the bones of habilines did from ancestral hominins. The Oldowan Industrial Complex, as it came (rather grandiloquently) to be called, extended to bones as well, which suggested that not only the skills necessary to create technology but that the *idea* of a tool could exist and could be carried from place to place along with suitable stones. Still, technology also showed gradations; similar if cruder artifacts could be found at *Australopithecus* sites. Raymond Dart, known for his work with the Taung child and the Makapansgat sites, argued that even australopithecines had an osteodontokeratic culture of reworked bones, teeth, and horns. If neither anatomy nor artifacts by themselves unequivocally indicated a new species, what did?[6]

There were doubters. Louis Leakey insisted there must be some older as yet unidentified ancestral stock buried deep in geological time from which "Dear Boy" had come and that constituted the true lineage for *Homo*. Further discoveries have tended to complicate rather than clarify; finds are both too sparse and too powerful. Each fossil find is unique, along with each narrative they force into being and each story of their discovery. Olduvai Hominin 8 (OH8) is a foot. OH13 ("Cinderella") is a cranium. OH24 ("Twiggy") was a hundred bone fragments. OH16 was trampled by cattle. Either the discoveries are pieces that can't simply be relegated either to variations or to new species, or they are massive (relatively speaking), such that each discovery seemingly causes the whole landscape of understanding to reorganize. The result is a kind of intellectual inflation in which too many ideas chase too few fossils.

There is a still larger question. It is not when does *Homo* first arrive, but when a question about early hominins gets answered. What constitutes an explanation? Scholars have turned to classification, analogy, and formula, all versions of types.

Nature is continuous, fluid, and ephemeral. It's the mind that imposes boundaries on time, space, cause, and species, that seeks fixed patterns in the flux, and that yearns to identify stable points of origin, determinative traits, and immutable referents. Words, ideas, images, and other representations are discrete, fixed, and once expressed, eternal. Numbers finesse the problem, because they can be infinitely subdivided and approximate some properties of nature better than words. Narratives, however, need anchor points for their spans; and context, in the form of comparisons and contrasts, requires stable sites or standards for reference. The common solution is to create types.

It's an old practice, among the oldest if we accept Alfred Whitehead's observation that Western philosophy amounts to a series of footnotes on Plato, for Plato made types into a transcendent presence. The "real" world, he argued, consists of ideals, unchanging and eternal, that impose form on matter. The material world, transient and flawed, is a poor approximation of that true world. Incorruptible knowledge—certain knowledge—is knowledge of the ideal. The classic example of an ideal is the circle. Our idea of a circle is perfect, while our efforts to draw circles are inevitably imperfect. The best way to know something is to abstract its ideal form from the muddle of marred and murky matter that makes up the world of everyday life. Plato's theory of Forms reconciled a messy and blurred world of data with a clear and invariant world of mind.

The modern mathematical solution is to replace ideal types with statistical averages, such that the mean value of a feature, drawn from many specimens, defines the standard for reference. But that requires a large sample size, and it demands that one or several traits serve as an index. Where data are sparse and each of the telling traits varies, or changes at different rates and times, the type dissolves into a hazy hive of buzzing information. The mind and narrative alike crave some immutable Form that unambiguously announces the origin of *Homo*, which is part of the appeal of Leakey's discovery. Anatomy alone can't do that, however, and instead of seamless continuities, the tendency is to present a chain of defined links. Inevitably, quarrels break out about the choice of the defining trait, the surety of the species identification, the confidence with which one might say that this specimen or this artifact is the product of this particular species.

In deciding to use the comingled tools as diagnostic, and to label the skeletal remains by their association with those artifacts, Louis Leakey leaped from bones to behavior and caused a stir that endures. Yet he did with fragments of data what others were doing by gluing bone fragments into a skull. He reassembled them into a coherent whole, a type, based on an ideal of what qualifies a species to be *Homo*. For such a task, you need borders; you need criteria for deciding whether some trait is in or out, whether it comes before or after the threshold. If anatomy doesn't suffice, then behavior, as manifest by percussing tools out of stones, might. Leakey himself doubted that the Olduvai find was the direct ancestor to modern humans, only a marker along a dendritic watershed whose human wellspring lay further back in the past; but announcing the composite of bone and stone as fundamentally human forced a discussion about what that ideal form of humanity might be and when it appeared.

Homo habilis became more than just a type specimen. It became a literal and metaphoric metric against which others could be compared and judged. As more discoveries stirred the dust, the essential attributes of the species, and that threshold of history where *Homo* segued from the other hominins, became more hazy. The original Olduvai specimen became less a true type than a holotype, the first of a cohort of specimens. Even more, perhaps, its discovery, identification, naming, and placement in the phylogeny of hominins made it an archetype of how the inquiry itself acts. The holotype said, This is what an emergent *Homo* looks like, and maybe what it acts like. The archetype said, This is what a discourse about such a discovery looks like, and how its discoverers act.

The Form of the Past

Even in the ancient world, critics recognized that Plato's theory, while useful for mystics and mathematicians, was itself an Ideal whose application to the ordinary world was as coarse and messy as its imagined realization of Forms into matter. It was especially tricky for the living world in which organisms grew and changed form. The task fell to Aristotle to devise a better model of explanation, one in which matter developed over time. It made sense then that he would also try to explain the character of narrative, and to exploit narrative as a mode of explanation. It is not enough to arrange data into chronologies. A narrative requires a plot that connects events with causes, or "movers."

This is a good approximation of what plagues so much of the discourse over hominin history in the Pleistocene. The quarrel over origins—just when the first *Homo* arose, exactly what trait in what proportion justifies classifying a relic as the founder of the lineage—depends on how we see the outcome. If we define hominin evolution (or *Homo sapiens*) by one trait, the narrative arc leaps one way; if by another trait, a different way. Today the trend is to focus on a suite of characteristics rather than one nominally dominant trait, but that suite will still seek to assemble itself into a recognizable type identified with a specimen, place, and story of discovery. Still, agreeing on origins remains complicated, not only because fossils are few, but because we cannot agree on who we are and why we are that way.

The constellation of proposed uniquely human characteristics holds many features, with more constantly offered. They typically take the form of humans are: the only creature who make stone tools; use fire; throw objects; use abstract reasoning; laugh; massively kill their own kind; or, as Mark Twain dryly quipped, "the only Animal that Blushes. Or needs to." In fact, each telling trait may not, by itself, be so special. Other animals are bipedal—birds since the Cretaceous, for example. Other animals use tools now and then. Australian eagles and Philippine tarsiers have been seen to pick up firebrands. Other creatures have large brains, are intelligent, or show self-awareness. Complex social organisms communicate among their groups. So it is, too, with anatomy. Over and again, as Richard Klein observes, "specialists cannot agree on which species and how many species existed at any one time," and when they can agree on a species, "there is still the problem of distinguishing between species similarities that were inherited from a common ancestor and those that may reflect only adaptation to similar circumstances."[7]

In this regard, *Homo habilis* may serve not only as a holotype specimen but a holotype of such controversies. Opinions of *Homo habilis* as paleoanthropologist Bernard Wood wryly notes, "differ." Some see the assorted fragments as part of "a variable, but well-defined taxon," while others believe the label has been applied "to such a heterogenous collection of material that the identity of the 'real' *H. habilis* is now all but obscured." Even the fixing of a canonical specimen has proved beyond the powers of science. Wood himself concluded that "answers" would only come when fieldwork had found many more samples and "deskbound analysts" had refined better their understanding of what variation meant. They had, in brief, to know more clearly what it was they were looking for and at. They needed a mental type.[8]

In the end, it is not the morphological form assembled out of fragments that may matter most but the Form in the mind that allows those fragments to become a whole. Even more, it is not *Homo*'s form but its function that most engrosses the modern mind, an insight that Leakey's leap from bones to tools captured perfectly.

Types of Knowing

New ideas and techniques appear from within disciplines, rarely from true hybridization from outside, and they grow as the discipline absorbs other ideas and refashions them in its own image. With regard to fossils, researchers will turn toward comparative anatomy and phylogenetic trees, the one to fill out the missing bones and the other to fill in the missing links. To explain how those fragments add up to humans, and to tell the story of discovery, they will turn toward narrative tropes, for which it is also possible to create comparative morphologies. Plato's forms may be only a shadow of their former selves, but they still fill caves, labs, and library shelves.

Richard Leakey once described the process by which workers find specimens in the field. There is a pattern, a fossil typology, that the surveying eye must match against the detritus and roughened exterior of the earth. "A fossil hunter needs sharp eyes and a keen search image, a mental template that subconsciously evaluates everything he sees in his search for telltale clues. A kind of mental radar works even if he isn't concentrating hard. A fossil mollusk expert has a mollusk search image. A fossil antelope expert has an antelope search image." That's how you find a bone, and how you find missing data for a paradigm, lost links in a phylogenetic tree, or character and event in a not-yet-completed narrative.[9]

Consider, next, how one piece might fit into an ensemble. From a single bone Victorian paleontologist William Buckland describes how to reconstruct a comprehensive skeleton. "At the voice of comparative anatomy, every bone, and fragment of a bone, resumed its place." Because each bone connected to the others in systematic ways, the whole was present in each part.

I cannot find words to express the pleasure I have in seeing, as I discovered one character, how all the consequences, which I predicted from it, were successively confirmed; the feet were found in accordance with the characters announced by the teeth; the teeth in harmony with those indicated beforehand by the feet; the bones of the legs and thighs, and every connecting portion of the extremities, were found set together precisely as I had arranged them, before my conjectures were verified by the discovery of the parts entire: in short, each species was, as it were, reconstructed from a single one of its component elements.

The power of the form triumphed over uncertainty and incompleteness and allowed the pieces to come together.[10]

There are plentiful counterexamples that describe what happens when the process has backfired as the idea becomes more real than the evidence, when the mind picks from its repertoire of forms figurations that it imposes wrongly over the scatter of data fragments exhumed from soil. In a famous episode, E. D. Cope reversed tail and head in assembling an *Elasmosaurus* (a plesiosaur), advertising his error not only in skeleton but in print, a mistake immediately recognized by his rival O. C. Marsh and subsequently affirmed by Joseph Leidy. (Cope never admitted the goof and never spoke to Marsh again.) In 1979, Noel Boaz published what he believed to be a hominin collarbone from a site called Sahabi in the Libyan desert—the site dated to five million years ago—and argued that the morphology indicated a bipedal species. At a conference in 1983, noted paleoanthropologist Tim White demonstrated that the "hominin" bone was the fossilized rib of a dolphin and took to calling the specimen "Flipperpithecus boazi."[11]

Still, the process is powerful, and it can expand beyond individual skeletons to whole genera or taxa. In such cases the unit is not comparative anatomy but comparative phylogeny; the species display a comparative evolutionary experience. The evolution of equids showed how one might expect the evolution of hominins to unfold; the spread of elephantids exhibits a pattern that erectines and sapiens echo; the survival and extinction of ursids suggests probable histories for Neanderthals and sapiens who shared similar habitats. With one part it is possible, cautiously, to reconstruct the whole.

Consider, finally, the formulas of narrative. Misia Landau has observed that the classic texts informing the history of paleoanthropology are "determined as much by traditional narrative frameworks as by material evidence." Such an insight moves into the comparative anatomy of story. Referring to Vladimir Propp's classic *Morphology of the Folktale*, she notes how much "[A]ll these paleoanthropological narratives approximate the structure of a hero tale." They narrate the story of "a humble hero who departs on a journey, receives essential equipment from a helper or donor figure, goes through tests and trans-formations, and finally arrives at a higher state." The form, she con-tinues, can accommodate "widely varying sequences of events, heroes and donors corresponding to the underlying evolutionary beliefs of their authors." Those who rely on such formulas are the Bucklands of narrative.[12]

But so are those who tell the stories of how human evolution has come to be understood, that is, the narrative of intellectual discovery. This is the story of those telling the stories of human evolution, and it also appeals to common tropes that create figured texts not out of bone fragments and lithics but from acts of insight and invention. Those moments leap from discovery to discovery, and connecting those caches into a coherent narrative is exactly the same process as constructing a phylogeny of early hominins. The narrative form endures, with a new set of actors. Instead of the bones of *Australopithecus afarensis* and *Homo habilis* and *Homo rudolfensis*, the story turns on the books of Ernst Haeckel and Elliott Smith and Theodosius Dobzhansky. Whether or not people can only know with reasonable assurance what it means to be human, they can know the origins and evolution of their inquiry.

Antecedents and Analogues

If comparative anatomy explains bones, comparative phylogeny helps explain evolutionary history. The species of hominins were many, and they shared a time and place with other megafauna. What happened to one genera might speak to the others as well; the missing links of history might be filled by analogy.

According to current understanding, the setting they all shared looked like this: As the Earth cooled for the first time in three hundred million years, yielding ice, sand, water, or wind according to local cir-cumstances, it twisted the handle of evolutionary ecology and smoth-ered, cracked open, spalled off, and otherwise upended landscapes. Old species went missing, new ones appeared, having, as Macbeth might have meditated, strutted and fretted their hour upon the stage. In Africa a presumed geographic unity further fractured. As forests broke up, a clade of hominins began to sieve, splinter, and spread, much as bovids and proboscids did. Along the Rift Valley habitats proliferated, filling up with niches like bric-a-brac in a Victorian drawing room. Amid this evolutionary ferment, the first *Homo* appeared.

Habilines and Hominins

Contemporary understanding holds that, when the Pleistocene opened, the ancestral hominins were a suite of australopithecines that researchers have sorted into categories of "gracile" and "robust" based on the specimen's morphology. They radiated across landscapes roiled by climate, crustal rifting, and competing creatures. None dominated everywhere. No single, unique morphological or behavioral characteristic made any one species universally successful. Southern Africa held *Australopithecus africanus,* an omnivore who consumed fruits, seeds, nuts, young leaves, and scavenged meat, and *Paranthropus robustus,* which favored a savanna salad of seasonal fruits, nuts, and seeds. Eastern Africa featured *Australopithecus afarensis* and *Paranthropus boisei,* along with such enigmatic relatives as *Australopithecus garhi* and *Paranthropus aethiopicus.* Both the gracile and the robust coexisted.

It was at this time—approximately 2.5 million to 1.6 million years ago in East Africa—that the first of a new clade, *Homo,* appeared. It arrives in the character of one patriarch, *Homo habilis,* and possibly another species, *Homo rudolfensis,* which might be a variant. But then almost everything about the founding species is unclear. How many others might have ranged across those flickering landscapes remains to be discovered and excavated out of tuff and limestone, and how they all relate will prove even more vexing.

The fact seems to be, there are too many and too few. There are too many specimens that look too different to arrange the lot into a simple phylogeny or narrative. And there are too few fossils—the sample is too small—to determine where the boundaries of each species lie. What range of variability might *Paranthropus aethiopicus* have? At what point among the fragments of *Homo* can we split the pool into *habilis* and *rudolfensis,* or should we lump them into one? Does *Paranthropus* really have three species, or are they merely two, or just one, or even a species of robust *Australopithecus*? They appear like lithic flakes scattered around a core. Which are discards and which intentional? If the point is to trace our modern humanity's lineage, which of the relic bones continue the line, and which are collateral? How much of the variety was simple ecological churn and how much genuine evolutionary turnover?

How exactly did the earliest species of *Homo* come from out of this swarm of australopithecines is phylogenetically unclear. Historically, a few paths entered the geologic woods and a few emerged, and how they connected within is, as yet, opaque. But roughly coincidental with the climatic upheaval that announces the Pleistocene, around 2.4 million years ago, *Homo habilis* arrived, and perhaps contemporaneously (more probably later), so did *Homo rudolfensis*. Until 1.7 million to 1.8 million years ago they shared Pleistocene Africa with each other and with *Paranthropus*. Both lineages were bipedal, had human (not apelike) teeth, possessed hands capable of finer motor skills, and lived in similar habitats, though in different regions of Africa. Each began not merely adapting to its environment but shaping it or, as the phrase goes, constructing a niche. The difference is that *Homo* could apparently ponder what it practiced.[13]

It's a crowded family album, likely to become more so as further archaeological discoveries ferret out shirttail second cousins and forgotten uncles from collateral lines. The extended family seems large because Africa is vast, and the onset of the Pleistocene had roughened its ecological texture; there were lots of places to disperse to. Meanwhile, what evolution gave, evolution also took away. The last appearance date for the australopithecines, both the gracile and the robust, was 1.2 million years ago. At this point the evolutionary plot begins to pivot around *Homo*.

This is, more or less, the consensus narrative. It has emerged, not unlike *Homo* itself, out of a welter of evolved stories, each tested against ever-changing circumstances. It gets rewritten with every discovery, and to help fill in the gaps, narratives can look outside the record of the hominins and beyond the logic of comparative anatomy to what might be termed comparative analogy, behind which might lie a type narrative.

Hominins and Elephantids

The saga of *Homo* was not unique within the Pleistocene for its evolutionary experience was characteristic of the times. What happened to hominins—a repeated proliferation of species within a setting of overturning environments—happened to clades of other taxa, as they contracted, collapsed or renewed, and spread outward. The climatic big chill led to a biotic big bang that dispersed new species widely. The similarities, however, extend to understanding as well. Those cognate creatures not only furnish models for organic evolution but serve as symbols that speak to humans in ways that illustrate their understanding of their place in the world.

Some creatures, such as the African bovids, share an evolutionary story of geographic radiation very similar to that of the hominins. The bovids, however, are so abundant that they become a kind of background noise to evolution. By contrast, the elephantids—the proboscideans—loom larger in storied potential not only because they closely shadow the hominins in their natural history but, more significantly, because they have interacted with cultural history. While the bovids might stand as the test case for the Turnover Pulse, the evolutionary history of the elephantids became a symbol for extinction and the might of the rising hominins. As such they have become an indispensable icon of the Pleistocene.[14]

The proboscids are an order of mammals whose Pleistocene evolution in almost eerie ways runs in tandem to, and at times intersects, that of the hominins. The order of Proboscidea is older, larger, and richer than the primates; three times they have radiated widely. Their diversity is made more bewilderingly complex by their ceaseless tendency toward convergence. Whatever their phylogenic point of departure, or even geography, "in spite of the remoteness of their common ancestry and the dissimilarity of their more immediate relatives," similar morphological patterns have evolved over and again. Like hominins they originated in Africa, and their fossils are found at the same sites.[15]

At the onset of the Pleistocene the elephantids, still lodged in Africa, saw two genera die out and perhaps six emerge, three or four of which led to different evolutionary lines. Like the hominins they moved from forest to more open landscapes; they speciated quickly; they spread rapidly. The dominant line was *Elephas*, which proliferated throughout Africa and then across Eurasia and through the Americas. The mastodon, the mammoth, and varieties of the Asian elephant were its primary expressions. Overall, it embraced between nine and thirteen species, which was comparable to the outburst of hominins. Some were giants, some were dwarfs. *Elephas* dominated in numbers, and *Mammuthus*, in migratory reach. As the Pleistocene pulsed, they thrived, while adapting, moving, and converging in form and behavior. Everywhere they became a dominant herbivore. Meanwhile, a stable but lesser genus, *Loxodonta*, apparently remained in what persisted of the African forest.[16]

As the Pleistocene ended, the elephantids underwent evolutionary pressures similar to those experienced by the hominins. *Elephas* began imploding, continent by continent, island by island, lost from Africa, from Europe, from the Middle East, from mainland Asia, from Japan, from Malta and Sicily, until only one species remained, *E. indicus,* sequestered in south Asia and isles like Sri Lanka and Sumatra. That vacuum left a huge empty niche in Africa that a single species, *L. africana*, quickly claimed by moving in from its ancient forest refugia. Save for *Loxodonta*, a narrative outlier, the story of the elephantids is much like that of the hominins and may in fact reflect not merely analogy but actions, for hominins and elephantids have come to share an evolutionary spiral that has led them to meet, often violently, on the steppes and woodlands of Africa, Asia, and the Americas.

That is the story as presently deduced from natural history. A parallel story exists, however, that traces the ways in which elephantids and hominins have interacted, and has given reason for modern humans to pause over the possible meanings such shared worlds might have and to ponder the ways in which their pasts might reflect on the present.

Some of the perceived identification comes also from the stuff of natural history. Both taxa are intelligent and are among a small clutch of species that includes dolphins, apes, and crows that show self-awareness, a form of consciousness. They are both capable of fantastic dexterity—an elephant can pick up a single strand of grass. They both communicate with sounds, gestures, and facial expressions, a kind of language. Both are social animals that function and learn within groups—migratory creatures capable of immense treks—and not merely keystone species but, perhaps, super-keystone species that dominate local faunal biomass and shape entire ecosystems. The elephantids emerged from the same biogeographic hearths as early hominins; they occupy some of the same cherished habitats and they exhibit many of the same dynamics of radiation and parallelism. Their paths have crossed repeatedly.[17] All of this explains a certain intellectual fascination with the elephantids, an identification as a kindred creature whose traits help elucidate ecological concepts and whose evolutionary experiences might help explain the saga of the hominins.

These are generic features, however, not unique to elephantids and humans. The specific bonding between the species strengthened as the Pleistocene unfolded. When the epoch spiraled upward, both groups spread and thrived. When it wound down, they came into lethal competition, and only two of the elephantids and two of the hominins survived into prehistoric times. And when modern science turned its gaze upon their relationship, the revelation of their mutual story became an important horizon in the history of ideas. In particular, the melancholy spectacle of vanished mammoths and mastodons stood as a critical moment in the intellectual unraveling of the Great Chain of Being—the type case for the reality of extinction. The study soon migrated from concepts of evolution into codes of ethics as a debate boiled over regarding the probable role of humans in directly or indirectly killing them off, for with their collision of destinies, the Pleistocene elephantids entered into the moral universe of human culture—a totem of human agency, a fable perhaps of humanity's flawed presence in the world, a cultural marker in moral evolution as significant as marker fossils are for geologic evolution.

Nor did this discovery, and its intellectual progeny, remain fossilized in the far past. The Pleistocene encounter with mammoths persists with the Anthropocene's ambiguous and often fatal contact with elephants. When environmentalists sought a poster species for the cause of wildlife protection, and to illustrate the moral turpitude of market hunting, they found it in the fate of modern elephantids and their ivory tusks. The African and Asian elephants have become as much symbols of contemporary conservation as the mammoth and mastodon for Pleistocene extinction. The elephant now stands as a veritable type for the ethical treatment of Earth and its creatures, as the descendants of the habilines—the handy man—seek to stay their hand rather than extend it further.

Origins: The Beginnings of the Ends

The need for stability amid the uncertainties—fixed geodetic points for seeing and understanding—is understandable. The quest for patterns appears at every level of research into human origins, but it takes a particular form for narrative, since end and origin must be linked. The usual strategy is to identify some distinctive trait and track its progression through time, from its onset to its present state (or demise). The instinct, that is, is to turn evolutionary opportunism into narrative surety and to stiffen phylogenic uncertainty into the crisp lines of story, like buried bone tissue hardening into limestone.

In reality, there are many traits, none of them diagnostic by itself; many fragments of many species, none of which easily aligns into a genealogical family tree; and many possible origins, none of them pointing to a singular moment of creation but to a slow crystalizing of characteristics. The promised surety of empirical science softens, like lithification in reverse. Each discovery not only forces a revision of the grand narrative by moving its site or date of origin but blurs that anchor point into a haze of statistics. With significant discoveries rare, each new revelation has the potential to restart and reconfigure the whole. An Ardi shifts the australopithecine narrative frame for Lucy. A Lucy compels a reexamination of how we got to *Homo habilis*. A Turkana Boy upends the story of development from *Homo habilis*. Such is the competition between the endless turmoil of empirical discovery and the required anchoring of narrative.

Science can sort through the empirical clutter and organize it into conceptual boxes and contribute data points to the story. Surely much of the awkwardness and idiosyncrasies of the involved disciplines will disappear as more evidence accrues. Larger swaths of the field will resemble a science, or at least behave more like one. But some of the peculiarities will persist, because the nature of the questions cannot be solved by more bones, artifacts, or lab tests since data sets lead to chronicles, not narratives. While science can construct plausible phylogenies, it cannot construct true stories, because, for purposes of narrative explanation, the end does not derive from the beginning, as science would have it, but the beginning from the end as Aristotle argued. The organizing principles for true narrative reside not in evidence but in rhetoric, and the themes that give it punch do not come from morphology but from a moral sense. Each mode of inquiry thus has not only its own sources but its own distinctive ways of organizing them. These are the types of narrative.

Fossil Bones: Deciphering the Geographic Narrative

Originally, the evidence for what the nineteenth century called "the Antiquity of Man" was a small cache of fossilized bone, and this remains the primary source material for the geographic narrative and has shaped the evolution of that discourse.

Like any body of evidence, bones and artifacts have their quirks and foibles, and so does a scholarship that must interrogate them.

The irrefutable nature of fossils is that they are tangible. What their shapes mean, and what their settings say, may be ambiguous, but the specimens themselves exist as material evidence. In the economy of knowledge they are land and bullion rather than services and promissory notes. They may be copied, in which form they can circulate as drawings and casts, but they cannot be manufactured or extracted by experiment. They are original, palpable, sui generis. They are not like abstract data from a laboratory trial: A researcher can't re-create them; they don't exist autonomously, as do numbers or concepts. Their discovery more resembles prospecting, and their excavation, mining. As with gold or gems, their rarity is what makes them valuable. And as with other hard-rock wealth, power resides with whoever actually finds or holds the specimen or controls the site.

By themselves the bones are mementos, curiosities, or objets d'art. They are the equivalent of holy relics or secular icons, and subject to pilgrimages like those to St. Foy in tenth-century England or King Tut in twentieth-century America. What bestows scientific meaning is their setting. Their context dates them, suggests the geography of the age they lived in, and identifies what other creatures and artifacts shared that time and place. But context applies also to their discovery and analysis. Much as there is a typology of bones, there is a typology of their sources, and to their finding, extracting, naming, reconstructing, and explaining.

Hominin bones come primarily from three kinds of contexts: gorges, caves, and quarries. Each of these sources has yielded type specimens; each has its peculiar style and methodology; and each sustains type stories of discovery. One story speaks to singular purpose. Like Heinrich Schliemann determined to locate the site of Homeric Troy, researchers identify a likely locale—a gorge or a cave system—and search with that dogged relentlessness that the Leakeys have displayed in the Rift Valley or Eugène Dubois did in Java. Most stories involve a mix of accident and system. They depend on someone doing something unrelated who finds something odd, and then gets it to someone who can make sense of it; this is the narrative formula followed for the first Neanderthals and the original australopithecines, and with the caves at Lascaux. Once a specimen is found, a site merits more methodical excavation. The fortuitous and the useful, accident and purpose—the process by which evolution becomes known seemingly mimics the dynamics of evolution itself.

The type locales differ in what they exhibit. The gorges—of which Olduvai Gorge in the Rift Valley is the best known—cleave through geologic strata and expose the bones. They reveal what would otherwise be hidden; in effect, nature digs an exploratory trench. They can stand for any site of erosion that leaves hard material on the surface. Discovery involves a syncopation between natural deposits, natural trenches, and a searching eye. The caves are principally depositories as they hold and preserve materials in a confined space. They have become as much a habitat for modern archaeologists as they were for early hominins. Think of such classic sites as Chauvet in southern France, the scene of spectacular cave art, and Sterkfontein in South Africa, where in 1947 Robert Broom and John Robinson discovered the australopithecine remains nicknamed "Mrs. Ples." Quarries are commercial sites for excavating stone, for which fossils are an incidental by-product. They do for archaeology what road cuts do for geology. They have yielded some of the field's most astonishing discoveries, along with its greatest hoaxes. The Kalahari escarpment yielded the Taung Child to Raymond Dart in 1925 and a gravel pit in Sussex led to the sorry spectacle of Piltdown Man in 1912.

The original bones were rare, rough-handled through geologic time, and fragmentary. The heavy work of analysis was done with casts, not unlike medieval scholars poring over copied texts. That proved true of ideas as well. Originals were difficult to find, and most scholars contented themselves with simulacra and glosses.

Fossil Genes: Decoding the Genomic Narrative

The story looks different when viewed from the genome, which appeals to its own distinctive data, relies on an alternative style of scientific inquiry, and not surprisingly constructs a somewhat dissimilar story of origins. Disciplines devoted to the genome diverge in what they regard as evidence, argument, and the informing principle of their implicit narrative. Sciences they may both be, but that is no guarantee that the geographic and genomic perspectives will see the world the same, or that they will express that vision other than binocularly.

The genomic narrative also has three main sources of data: proteins, nuclear DNA, and mitochrondrial DNA. Unlike fossils, they are available universally since each human being today contains the entire genetic history of the species. Among genomes there are variations sufficient to account for the dispersion of *Homo sapiens*. There are differences attributable to geography and history, but there are no issues with having to find bone chips or smashed cranial plates that might or might not be at a particular site. The potential data set is unbounded: the billions of humans who live today. The information was and is everywhere; it can be reproduced by everyone; and it can be confirmed in labs. Yet no less than with fossil bones, there remains a story of first discovery and analysis, often infused by cultural and nationalistic biases over who descends from whom.

The genomic narrative is an evolutionary chronicle that follows from the crude genetic material of the primates as shaped by natural selection into present-day hominins. What constitutes a new species depends on how much the genomes diverge, and when this happens it can be dated by the rate of mutations. The construction of phylogenetic trees is a matter of statistics and cladistics. The record over time is in fact a biochronometer with the rate of mutations resembling a biotic analogue of radioactive decay. Like genes, the timepiece does not depend on geographic setting and the whims of deposition and preservation. The story of discovery follows the templates of laboratory science.

Still, similarities abound between the two dominant narratives, among them the character of the conflict that results when they disagree. The quarrel between field and lab is old and comes with its own internal narrative. The usual assumption is that when hard and soft sciences clash, the hard science must win. The laboratory trumps the field, and what isn't numbers isn't science as Lord Kelvin famously sneered. Numbers, though, can be as anecdotal and ambiguous as bones. After all, Kelvin denounced Darwinian evolution because the Earth, as then conceived by physical theory, could not be as old as Darwin's theory said it had to be. Assuming the planet had cooled from a ball of incandescent gases, its rate of cooling could be determined in the lab, and the age of the Earth could be calculated with mathematical rigor. It was a matter of simple arithmetic—Darwin had to be wrong. In fact, Kelvin was wrong. His calculations were only as good as his theory and the geologic evidence behind it. When a new source of heat was discovered in the form of radioactive decay, his numbers became as useful as yesterday's losing Powerball ticket.

Fossil Texts: Deconstructing the Cultural Narrative

Such intramural squabbles pale, however, before the larger contest between the sciences and the humanities. Here, the assumption is, once again, that the tough-minded will quash the tender-minded, that numbers will overwhelm figures of speech, that the anecdotal humanities will merely decorate the fundamental text of the sciences, like vignettes on an illuminated manuscript.

Yet while bones may be tangible fact, "what they tell us," as David Pilbeam notes, "is highly ambiguous. Interpretations, schemes, and stories vary from one authority to the other, and evolve and reemerge through time." The facts chosen, the preferred methods and technologies of analysis, the cultural context within which they absorb meaning—all depend "upon theoretical background assumptions, many of which are either not acknowledged or not even recognized."[18] They percolate through the culture at large. Histories emulate the forms of narrative typical of an age; values echo from one cultural cliff to another. As Lionel Tiger has wryly observed, the meanings imputed to the newly reconstructed Ardi (*Ardipithecus ramidus*) reflect the "moral coding" of contemporary values, such that the evidence from teeth and bipedalism seems to lead to "the affable males and comfy families of Berkeley, Calif., or Ann Arbor, Mich."

He continues, "How convenient that our evolution took this correct pro-social form." In brief, confronted with immense ambiguities, analysis reverts to type, in this case, to the tropes and templates of those arts and scholarships that must deal openly with values.[19]

The sources here are primarily texts and the ways to analyze them—the usual stuff of the humanities. This scholarship, too, has its characteristic evidence, organizing schemas, and its peculiar pathologies. At stake is something more than how chronologies of hominins get rewritten into narratives that invest them with cultural meaning. The issue is how the discoveries and interpretations are described, how we understand the larger enterprise, how that scholarship itself behaves, and how its sustaining society engages with it. In its abstract form it is the story of how the stories get told.

There are some eerie intellectual echoes at work. Medieval Europe began its reeducation by recovering texts buried in antiquity. These were the fossils of their times; acquirers and translators, the excavators; and professors, those empowered to hold forth about their meaning.

The text was the essential unit of discovery and analysis. Texts were rare, they required often laborious translations, they spread by copies. He who held the text held power. Students paid to be lectured to— that is, read to from the text. In the process of transcribing, they acquired a text of their own (call it a cast). The texts were scattered and fragmentary; they had to be reconstructed and connected. This required painstaking commentary, termed glosses, and the glosses were transcribed between the lines of the original work. Someone who merely glossed over a subject or read between the lines confounded the subject. They made coherent—according to their organon and view of the world—what existed in the sources only as textual shards.

As bad, the texts were recovered in random order. Texts written early might be found and translated after a later one. A book by Aristotle might become known before one by Plato that it comments on, a work of Archimedes before one by Euclid from which it derived, a play by Euripedes before Aristophanes. Scholars sought to assemble the growing corpus into something like a coherent system of knowledge. They had to construct a phylogeny (as it were) of the received texts. Because Aristotle was recovered relatively early, and in large chunks, he became, as Dante put it, the "master of those who know," and his philosophy, reconstructed from the fragments, became the basis for the great medieval *summas* of Aquinas and his cohort in the twelfth and thirteenth centuries. That accident—the happenstance of sources and their discovery—shifted the evolution of learning along a particular path. Had the complete dialogues of Plato been found first, philosophy would have taken a different direction, or had the corpus of Archimedes been found before Plato, Western learning might have taken an alternate evolutionary path altogether.

The crunch came when the desire for more texts exceeded the supply. There were more and more scholars, and greater and greater demands placed upon them, and the result was a proliferation of glosses. Scholarship slipped into scholasticism. Commentaries increased faster than original texts, and then came glosses on glosses, and the whole enterprise threatened to become an increasingly ornate and elevated construction on ever shakier foundations. Eventually the Renaissance scorned the edifice as a medieval Tower of Babel. Its scholars wanted the originals, which they would translate themselves. *Ad fontes*—to the sources—was the rallying cry. Add to that project what Francis Bacon called the *novum organum*, or new logic, of what became modern science, in which it became possible to read the book of nature as well as the books of the ancients. In some respects, for the study of human origins, the discovery of genomic data has served as a *novum organum* that has helped to break up the hegemony and hoarding that goes with the limited bones of discovery.

An estimated 210 skulls of paleohumans exist, but there are hundreds of paleoanthropologists. It is harder to discover new skulls (and skeletons) than to create new scholars to examine them. At the Paleoanthropological Society's 2011 meeting approximately 273 participants delivered fifty posters and forty-six podium talks; the society maintains a mailing list of 1,300.[20] No one should be surprised if, as evidence, fossils seem to resemble the texts of the humanities and if nominally scientific discourse should veer into scholasticism.

So if the discoverer of a critical fossil resembles in some respects a mining magnate, he also resembles a scholar of medieval times. To hold the text is to hold the tangible source of evidence. Rare books are treated like rare fossils—housed in special casings, subject to strict usage. The Dead Sea Scrolls, discovered in 1947, were, until the 1990s, effectively a monopoly of those researchers who had physical possession; this was particularly true of fragments. Such behavior is not, in principle, how modern science is supposed to behave, but it is how other forms of scholarship have behaved in the past, and some continue to do so. A major reason seems to be the nature of the sources. If they are ample and open, their study behaves one way. If rare or hoarded, they behave another.

Chapter 5

Out of Africa

There is always something new coming out of Africa.

—ancient Greek proverb

FOR INTELLECTUAL AS WELL AS natural history, *Homo erectus* is the linchpin hominin of the Pleistocene. The erectines were the first known hominin to boost cranial capacity significantly relative to body size, the first to possess a complex toolkit, including hand axes and torches, the first to leave Africa, and the first to migrate across whole continents, and they are—by far—the longest lived. They confirmed hominin evolution to science. They made real, for the first time, the hypothetical argument that natural selection might apply to hominins. Their story is the story of some two million years of Pleistocene history.

Yet it is a story that is tricky to tell. For one thing, the narrative movement shifts from species evolution to species migration. The erectines are famous for *not* changing anatomically for nearly two million years; instead, they carried their inherited form out of Africa and spread over most of southern Asia. It is no longer enough to connect fragments historically through phylogenies of descent. Rather, the pieces have to be joined geographically. In odd ways, the genomic narrative limps along while the geographic narrative races ahead. For another, the erectines do not offer dramatic turns of plot. Unlike the habilines, they do not introduce the epoch, and unlike the sapiens, they don't conclude it. Rather they simply enlarge its scope. They are simply there, like bovids. Or as one wit quipped, they are the Greek chorus of the hominin Pleistocene.

Their story also shapes the story of understanding them. Comparative anatomy and comparative biogeography had themselves histories; the diffusion of hominins beyond Africa inspired a diffusion of research traditions and centers. Whether or not *Home erectus* speciated due to geographic isolation, the study of human origins did. Different research styles and themes emerged and created the intellectual equivalent of new species such that vital information could no longer exchange freely between them. Every new find seemed to create a new species, each with its distinctive narrative. (The erectine Java Man, discovered in 1891, was not originally placed into *Homo* but became its own genus, *Pithecanthropus erectus*; a 1949 discovery by John Robinson did the same, beginning as *Telanthropus capensis* and evolving into an erectine, or perhaps into *Homo ergaster*.) A universal story splintered into regional ones, which tacked close to nationalistic sentiments. The historical geography of hominins led to a comparable phylogeny of explanations. Eventually the fundamental unity of the far-trekking erectines reasserted itself, and a consensus narrative emerged that identified human origins with Africa.

Yet as a character *Homo erectus* strides somewhat awkwardly across the Pleistocene's narrative. His is the literary problem of a second act in a three-act drama. While ancestral, he is not even what paleoanthropologists and archaeologists term archaically modern. No one would mistake him for an australopithecine, but neither would they mistake him for a Neanderthal, much less a contemporary modern on a commuter train. He is too recent to radiate an aura of the hoary and the mystique of origins and too old to join the ferocious squabbling over the often negligible differences by which authorities debate degrees of separation between anatomically and behaviorally modern humans. The erectines, as one authority has put it, were not "'trying' to be human. . . . They were very good at being *H. erectus*, and that was enough." If longevity is the definition of success, "then we have a clear winner in human evolution, and it is not us."[1]

The erectine story spans virtually the whole of the Pleistocene, but not quite. Fascination most fixes on the times of sudden change, and particularly on those moments (and species) that define beginnings and ends. The erectines are not party to those contentious debates about when the Pleistocene commences and when it concludes. They instead bind together the vast center. The Pleistocene's hominin story is largely theirs.

Parallel Paths of the Past

When the erectines appeared around two million years ago, the Pleistocene continued much as it had before, with kaleidoscopic mixings of air, earth, water, and life. When the circumstances were favorable, creatures could move over vast terrains, and match that geographic dispersion with a genomic one as new species flaked off from the core stock. Timing and location varied as terrain and climate opened and shuttered potential routes of transit. Mountain passes became passable as ice melted; saltwater straits dried into plains when sea levels dropped; isthmuses became supporting thoroughfares as rains turned deserts into grasslands. This happened for many taxa that arose in one place, migrated and mutated, and eventually survived or even flourished in places far removed from their point of origin.

The hominins were part of this vast cavalcade. The story of their rise, dispersion, and extinction matches those of the antelopes, elephants, horses, bears, and others with whom they shared time and place. Their value—added for narrative—derives from the fact that all intersect with modern humans. The bovids, particularly antelope, are a robust but otherwise undistinguished group whose story establishes a baseline for concepts of environmental change. The proboscids, or the elephants, became a symbol of the Pleistocene and their extinction a symbol of what the rise of humans has meant. The equids—horses—emerged as an emblem of evolution itself, such that their phylogeny became the type for any group that might emerge, spread, and sink. Instead of a morality tale like the proboscids, the equids told a fable of ideas about how concepts can arise and transform over time. The ursids, or bears, so similar in habits and habitats to Pleistocene humans, provide an evolutionary Other, an alternative ending to an oft-shared narrative. In their various ways these histories demonstrate that in evolution, unlike in Euclidean geometry, parallel pathways can meet.

Erectine Exodus

The earliest hominin fossils other than *Homo sapiens* were discovered almost three years before Darwin's *Origin of Species* was published. They were found in Europe, where interest and scholarship resided, and they were the most proximate (and therefore most abundant) of relatives, what became known as Neanderthal Man. Still, there was more than a little historical contingency involved, since the fossil came with no dating context. The bones were discovered as curiosities, objects collected and exhibited as part of the bric-a-brac of natural history.

An understanding of how intermediate types might be aligned into a family tree came from the fossils of other, more abundant species, notably the equids. Within a decade the evolutionary history of the fossil horse established a type phylogeny, a model of development (or "descent") that could be applied to other organisms. It furnished a template by which researchers interested in human origins could move from a common ancestor of apes and humans to contemporary people. In 1879 Ernst Haeckel drew the first ancestral tree (as distinct from genealogical charts) in which he showed a trunk rising from Apes to Ape-Men to Man, with various categories of apes branching off. That suggested clearly what the intervening forms ought to look like. Haeckel went further and specified what he regarded as the intermediate form ought to include. While fossil equids had confirmed the theory of evolution, the theory, now accepted, directed the search for fossil hominins.

But where to find them? The prevailing consensus assumed that they would be found, like Neanderthals, in those regions where the grandest civilizations flourished, that is, Europe or Asia. Eugène

Dubois, a Dutch anatomist, thought they would be in the East Indies, where the orangutan and gibbon are found and where, conveniently for his fossil hunting, the Netherlands had colonies. In 1887 he joined the Medical Corps of the Dutch East-Indian Army, settled in Java, and commenced the search for what he overtly proclaimed would be the Missing Link. Incredibly, between 1890 and 1892, he found it with a fragment of mandible, two teeth, a femur, and a cranial cap. Since the biggest distinction—or at least the first point of segregation—between apes and humans was an upright posture, which this creature had, it was dubbed Java Man, and later identified as a type of *Homo erectus.*

In 1921 additional hominin fossils were found at Zhoukoudian, China. Over the years they appeared throughout the African Rift Valley and in Dmanisi, Georgia. The earliest expression, in Africa around two million years ago, has archaic features that some paleoanthropologists believe warrant a separate species, *Homo ergaster;* a later expression (circa 800,000 years ago), arising again in Africa, some believe should also qualify as a distinctive species, *Homo heidelbergensis.* The ergasters stayed in Africa; the heidelbergentines migrated into Europe; the rest diffused throughout southern Eurasia. By the end of the twentieth century, some 150 specimens had been discovered; granted their "(currently intractable) taxonomic issues," not a few researchers are inclined to lump them all together as *Homo erectus.* They differ only by "relatively subtle cranial characteristics," notes Richard Klein, which makes nominal speciation a matter of "geography and dating" rather than "morphology."[2]

The new arrival soon left, and the newcomer eventually became an old-timer. No previous hominin had wandered so widely, and none had lived so long. The earliest fossil, from the Rift Valley, dates from 1.8 million years ago; the last, from Java, perhaps lasted until 50,000 years ago. Dates suggest that the erectines flourished in Asia from 1.2 million to 1.7 million years ago. What may be most curious is not that up to three closely related species might have arisen in Africa and dispersed differently, but that, having dispersed for a million and a half years, more speciation did not occur.

Although the erectines were world travelers, their world was one cir-
cumscribed between what they could do and what the Pleistocene
might allow. Bridges and barriers were relative, since a strait might
not stop a creature who could build a raft, nor cold a primate who
could carry fire. The choke points were several, and they were met
sequentially. The first involved egress from Africa. Until 1.5 million
years ago there was still passage across Aden that joined the African
Rift to the Arabian Peninsula; then it ruptured. That left a gate at
Sinai: when dry, a dam, and when lush, a sluice. What paleoclima-
tologists have termed the Saharan pump, the climatic pulsing that
turned the land from desert to savanna and back again, pushed and
pulled species across the Sinai. The erectines were the first hominins
to make the passage, an erstwhile exodus.

   Then they moved east. The Levant became, then as now, a cross-
roads. To push into Europe they had to force passage over the
Caucasus Mountains or the straits at the Bosporus and Hellespont,
and among the curiosities surrounding the erectines, otherwise so
far-trekking, was that they did stride over now-submerged coastal
plains. They made that passage late, perhaps as a subsequent subspecies. Many simply stayed, exchanging Africa's Rift Valley for another
in the Levant, and others continued to Asia. Once in southeast Asia,
they confronted two geographic gates, one to the northeast and one
to the southeast. Each swung from barrier to bridge, as interglacials
raised sea levels and glacials lowered them. The northeast, known as
Beringia, led to North America. The southeast led into Sahul, as the
welded continent of New Guinea and Australia became known. The
barriers to the north apparently proved too formidable, and erectines
probably never made it to Beringia. The barriers to the south were
more porous.

   The erectines were not, it seems, cold adapted; they hugged south
Asia, even out to Indonesia, until blocked by the Lombok Strait,
which was laced with deep and treacherous currents even during
lowered sea levels, that demarked the Wallace Line, long recognized
as among the most powerful faunal filters on Earth. Still there is

evidence that erectines did reach islands within the Mediterranean (Sardinia and Crete) and East Asia (Okinawa and Timor). A most intriguing prospect focuses on Flores, a small Indonesian isle east of the Wallace Line that may have been occupied since 850,000 years ago. If confirmed, such seafaring would require "a virtual re-writing of our assumptions about the [erectines'] cognitive, technological, social and perhaps intellectual capabilities," as R. G. Bednarik notes.[3] But that has been the curious task of *Homo erectus* from its original discovery. Somehow an early hominin did reach Flores, and almost certainly it was an erectine.

The erectines were the first hominins to radiate widely out of Africa. Others followed the path they blazed, and then pushed beyond. If the Pleistocene begins with the climatic aftershocks that followed from the sealing of the Isthmus of Panama, it might be said to end when a descendant of the wanderlusting erectines excavated a passage through it.

Equid Exemplar

In 1876 T. H. Huxley visited O. C. Marsh to view the Peabody Museum's collection of horse fossils. Marsh doodled what the ancestral horse might look like. Together they speculated about how the evolutionary histories of equids and hominins might compare. It was an epiphanic experience from which Huxley concluded that the gathered specimens "demonstrated the evolution of the horse beyond question, and for the first time indicated the direct line of descent of an existing animal." In the saga of the horse, evolution had found its exemplar. "The line of descent appears to have been direct," Marsh intoned, "and the remains now known supply every important intermediate form." The fossil horse was abundant and accessible on several continents, and its discovery was well-timed with the emergence of evolutionary theory.[4]

Over the coming decades the major minds of paleontology, biogeography, and evolution repeatedly reached for the chronicled evidence of equid teeth and hooves to elucidate one biological "law" after another, and to find confirming proof for species radiation, extinction, and descent. The evolution of the horse became a type, the template against which the stories of other genera might be compared. It demonstrated how to work through fossil data and what the result would look like. "The history of the horse family," as G. G. Simpson affirmed, "is still one of the clearest and most convincing for showing that organisms really have evolved, for demonstrating that, so to speak, an onion can turn into a lily." That history showed, too, how migration, a movement over space, could reconcile with evolution, a change over time.[5]

As with many mammals, the ancestral stock of the equids underwent an extensive radiation during the Miocene, and then got squeezed during the Pliocene and entered the Pleistocene with two or three major clades. Over and again a rhythm of renewal, dispersion, and collapse replayed itself. The center for each dispersion was North America, and, with each outburst, the geographic range of equids expanded. In the Eocene equids went from North America to Eurasia; in the Miocene, they penetrated into Africa; by the Pleistocene, with the Isthmus of Panama open to traffic, they spread into South America. During the Pleistocene they reached their maximum extent, failing only to cross the Wallace Line into Australasia. As the epoch ended, however, *Equus* imploded. It became extinct in its ancestral American hearth; it survived, barely, in central Asia and in Africa as zebras and asses. The survivors experienced another radiation, this time as domesticated livestock.

The Pleistocene phylogeny of *Equus* arguably parallels *Homo*. It begins with a small stock, speciates widely, and then collapses into one or a few survivors that persist because of their relationship to the sapiens. By the Pleistocene it was possible, at least episodically, to trek over both Old and New Worlds; only remoter islands, New Guinea, and Australia were removed from routine transit, and Antarctica alone was utterly isolated from terrestrial traffic. In this economy of exchanges, Africa was at times a source and at times a sink.[6]

Such natural histories apply to many genera, however. What makes the equids especially attractive is what also makes the proboscideans interesting: their intellectual history. Within a decade after Darwin published *On the Origin of Species*, the fossil horse seemed to demonstrate its arguments with paradigmatic rigor, a full description of descent. Within three years after Darwin extended evolution by natural selection to humanity with *The Descent of Man*, the exemplary phylogeny of the horse suggested what the intermediate forms of hominin evolution might look like and how they might be assembled into a whole. Like *Equus*, its researchers moved and merged between continents. The prevailing belief that evolution was progressive (or "orthogenetic") found perfect expression in the advance from tiny *Eohippus* to majestic *Equus*, as species after species seemed to move directionally from small to big and from simple to complex.

The story of its discoveries, and disputes over priorities and interpretations, provided a narrative template for paleontological pursuits generally. Whether or not the horse was a paradigm for evolution, the reconstruction of its phylogeny became a paradigm for how to do evolutionary science and what output one might expect. The pattern quickly translated into hominins no less than fossil hippos and dinosaurs and was applied to paleoanthropologists no less than to prospectors for the rare bones of an *Archeopteryx* or *Mammuthus*. The public quarrel—at once comic and disgraceful—between E. D. Cope and O. C. Marsh foreshadows generations of bickering, hoarding, and posturing by successors for whom a rare discovery might bring singular fame and fortune.

Not least, the horse was more than intellectually useful; it was lucky. Caches of bones could be found near the major centers of research, which is to say, Europe and North America. Marsh found *Eohippus* by having a Union Pacific train stop at the Antelope Station in Nebraska, long enough for him to inspect a well from which, reportedly, early human bones had been unearthed. Instead he found "many fragments and a number of entire bones, not of man, but of horses diminutive indeed, but true ancestors." *Equus* was lucky, too, in its ability to attract the major figures of the field. A founder of comparative anatomy, Richard Owen, worked on European horse fossils; the patriarch of vertebrate paleontology in America, Joseph Leidy, published his inaugural scientific paper on Pleistocene *Equus;* the first American professor of paleontology, O. C. Marsh, assembled the phylogeny of the equids; the great propagandist of Darwinism, T. H. Huxley, transformed that project into an exemplar of evolution by natural selection; the man who integrated paleontology into a modern synthesis, G. G. Simpson, used the horse to demonstrate neo-Darwinism as Huxley and Marsh had with its progenitor. Even today, as paleontologist Bruce MacFadden notes, "fossil horses are at least on a par with fossil humans and humanlike ancestors as examples of evolutionary patterns and processes."[7]

As evidence mounted that descent was neither simple nor singular but messy, and that progress was an illusion, the horse retained its intellectual standing. And it also served as a model for the evolution of science; its interplay of experiment and serendipity, of field and lab, of personality and institutions. The history of its explication demonstrates that the descent of science is also as populated with false leads and failed trials as nature, and that only in retrospect does the chronicle seem providential.

Timing was (almost) everything. *Equus* was the right genus at the right time, and its significance galloped across the emerging intellectual landscape of early evolutionary thinking as its multiplying species had exploded across the Pleistocene. If Marsh's doodlings of *Eohippus* are later variants of the Lascaux and Altamira cave paintings, in which horses were prominent, his diagrammed pedigree, as a visual metaphor, restated an ancient relationship between equids and hominins. The image became almost Platonic in its power to impose itself on the awkward matter of actual bones. Etched on paper and into mind, the pictograph became itself a diagnostic fossil in the history of evolutionary thought.

Erectus Unbound: Quest for Fire

Its toolkit did not, at first, much distinguish *Homo erectus* from its predecessors. The collection of stone choppers, scrapers, and flakes that the habilines possessed, now known as an Acheulean assemblage, remained remarkably unaffected by the appearance of the erectines. It was not a new hominin technology but where the new hominin took the old that mattered. A major change did occur around 500,000 to 600,000 years ago, when choppers acquired wooden handles and became hand axes. This technological breakthrough occurred roughly with the appearance of *Homo heidelbergensis*.

Such tools extended the physical capabilities of early hominins. But they were replicas and compensations for otherwise meager faculties. Hard-edged stone substituted for talons, teeth, and claws. Axes added heft to limbs and power to muscle. Heavy choppers could smash what lighter mandibles could not. Throwing objects offset poor leaping. Such tools imitated anatomical features. They endowed hominins with abilities that, for other creatures, natural selection knapped out of genetic cores.

But one technology represented a change in kind. *Homo erectus* would handle fire: could certainly get it; could perhaps make it; could clearly tend it; and could unquestionably blend it with other arts. Fire became the universal catalyst for almost all human technologies, the essence, according to universal myths, of what most distinguishes humanity from the rest of creation, a paradigm for domestication, and a master metaphor. In *Ecce Homo*, Friedrich Nietzsche declares:

> I know whence I originate!
> Like a flame insatiate . . .
> Surely, flame is what I am!

Even today fire continues to stand for what is uniquely human—traits not cached as fossils, a technology of process and behavior, not one of material artifacts.

Carrying the Fire

Fire is the oddity among ancient technologies, for it is not a material
object but a chemical reaction. You don't carry fire, you carry the
conditions that permit it to happen; you can't make it and set it aside
for later use, you have to tend it. You can't find fossil evidence of fire
itself, like a lithic flake or drilled bone; you find its chemical residues,
such as char and ash. So, too, its effect on early hominins, while powerful, must be manifested indirectly.

No one invented fire—it has been on Earth for 420 million years.
It thrives best under a climatic order that features rhythms of well-
defined wetting and drying sparked by lightning. Conditions have to
be wet enough to grow fuels, dry enough to burn them, and subject
enough to sparks to take flame. Such a regimen describes almost
perfectly the circumstances of the African savanna and the Rift Valley. The spread of C_4 subtropical grasses embedded fire even further
into the regional ecology. The putative hearth of humanity is in truth
a natural hearth. Early hominins would have grown up amid fires as
they did antelopes and leopards. They would have scavenged among
its burns and occasional kills. At some point they would have picked
up a brand or gathered embers as they might a bone or flake. It is
hard to imagine any scenario other than accident that led to its first
use. The insight would have been that it could be perpetuated, and
for this it is hard to conceive any model other than tending children.

The technologies of fire begin with making it. The simplest techniques involve casting sparks or briskly rubbing, eventually by drills.
Undoubtedly the skill developed as a by-product of other toolmaking, but there is little evidence that erectines used fast scrapers or drills.
They did strike stones, and eventually flint would shower sparks, and
the sparks would land in tinder, and the toolmaker became a fire
maker. He would carry suitable fire stones as he did lithic cores. Or,
as likely, he would simply carry the fire, as Australian Aborigines and
Andaman Islanders and endless others have. The identification of fire
with life is surely prehistoric, and not altogether invalid. Fire, though
not alive, is a creation of the living world and displays many of

its outward properties. Life supplies the oxygen it needs to breathe; life furnishes the combustibles that feed it. Like living creatures, it warms, it moves, it sounds. It is birthed—in many languages (as in English) "to kindle" applies equally to giving birth as to starting a fire. It must be tended. It must be bred and trained. When left, it is buried. Carrying fire, that is, involves not simply a chemical reaction but a social one. It demands a reorganization of tasks and duties. And not least, fire has to be housed. Paradoxically, a fire has a greater need for shelter than does its tenders. This—creating a *domus*—may well mark the onset of domestication. Particularly if fire was onerous to start, keeping it always alight required shielding it from the elements. The instinct for eternal flames is likely encoded deeply in hominin culture.

Hearth and Home

Controlled fire brought light and heat, but direct application most likely commenced with cooking, which remains the paradigm for pyrotechnologies generally. What began with meat and tubers eventually fed bone, stone, sand, metal, liquids, wood, whatever might be found, into the transmuting flames.

Heating potential foodstuffs improves their biological payoff dramatically. It converts raw biomass into physiological fuel. It denatures protein, gelatinizes starches, and otherwise makes raw biomass more digestible. It transforms barely edible starches into higher caloric carbohydrates. It detoxifies foods of many harmful chemicals and kills off worms, bacteria, and other disease bearers and parasites. For a given harvest of foods, cooking makes eating it easier and multiplies its nutritional values.[8]

Simply by cooking food early hominins expanded the range and richness of their diet. The alternative has returned in modern times with "raw food" cultists for whom cooking—long an emblem of civilization—consider the fire the origin of those unnatural processings that have rendered so much of contemporary food unhealthy. The outcome? Even with access to fresh (if raw) foods year round, and to nutritional supplements, researchers have concluded that "a strict raw food diet cannot guarantee an adequate energy supply." Cooked tubers are more potent than raw meat. Myths about the origin of fire attribute uncooked food as the most savage of deprivations.[9]

There was a change in human morphology that reflected a change in diet. A shift to more meat is likely a partial explanation, but cooking is a more wide-ranging and profound one. It could affect everything. Cooking was a change not imposed by external forces, but by an internal insight. No outside factor compelled hominins to cook what they ate. Certainly they would have found heated food from time to time on a naturally combusting savanna. But using fire was a choice. With the adoption of a cooking culture, the hominins began to tinker with the evidence about themselves that would be available to their inquiring descendants.

Some changes belong in the digestive tract. We have, as Richard Wrangham notes, compared to our near relatives, "small mouths, weak jaws, small teeth, small stomachs, small colons, and small guts overall." All these means to break down food physically and chemically can shrink because cooking has already begun the process. The most auspicious reforms, however, focus on the skull. No longer required to chew tough matter, the teeth and jaw can decrease, and the musculature needed to power them become less massive. That means the skull need not act as an anchor post to hold them; it can lighten and bulge upward without dietary and selective cost. It can accommodate a much larger brain.[10]

The upshot was the largest leap in anatomical change recorded among the hominins: the most dramatic change in tooth size in six million years; the largest boost in body size; the hugest increase in cranial size (42 percent); and a sudden capacity to trek over whole continents. Something happened to endow *Homo erectus* with capabilities that restructured anatomy, physiology, and behavior in ways not easily deduced from environmental upheaval alone. Since then, while cranial capacity has increased, no other reconstruction of morphology has occurred on a comparable scale.[11]

It seems improbable that no large changes followed from routine cooking, which would be the case if *Homo erectus* acquired fire during its nearly two-million-year history. The oldest hearths, evidence for fire tending, date to 1.42 million and 1.6 million years ago. The oldest archaeological evidence for fire *making* is 790,000 years ago in the Dead Sea rift valley. By then the basis for fire technology was fully realized, and granted the conservatism of Acheulean technology generally, it is probable that erectines had it from the beginning.[12]

Pyric Paradigm

Cooking became the pyrotechnic paradigm for other uses of controlled combustion. Fire granted light, power, warmth. It could be applied to stone, wood, sand, metal, liquids, air—all the other elements of the ancients. One could cook wood to harden it into spear points; cook stone to soften it for easier flaking; and, over millennia, learn to cook rocks to crack open for mines; to melt ore into metals and sand into glass; to transform clay into ceramics; and to fell and hollow trees. Almost any reaction was quickened and strengthened by heating. Controlled fire was the essence of alchemy, the foundation of modern chemistry, and, as Aeschylus has Prometheus assert, a universal catalyst for all the other arts of humanity.

Of the matter reworked by flame, raw humanity was itself among the first. Fire altered his anatomy; cooking changed diets enough to allow the skull to swell and the digestive tract to shrink. It altered behavior; he had a duty to tend fire and apply it wisely. The hearth became the focus for social life; around its flames, cooking took place, the band warmed, and, with language in its toolkit, the clan told the stories that passed on its practical lore and sense of itself. They carried their fire with them. They identified with it. Other tools could substitute for the talons and senses of fellow creatures, creating a degree of parity. Fire placed hominins apart.

The recognition that fire lies at the basis of technology, and hence of humanity's power, did not end with Aeschylus. In his *Natural History*, Pliny the Elder marveled how "fire is necessary for almost every operation. . . . Fire is the immeasurable, uncontrollable element, concerning which it is hard to say whether it consumes more or produces more." In his summary of Renaissance lore, *Pirotechnia*, Vannoccio Biringuccio observes that almost all technology was pyro-technology, because it depends "on the action and virtues of fires." In 1720, well into the scientific revolution, Hermann Boerhaave announced that "if you make a mistake in your exposition of the Nature of Fire, your error will spread into all the branches of physics, and this is because, in all natural production, Fire . . . is always the chief agent." And the Industrial Revolution bestowed power by applying a technology from a distant past to burn fuels extracted from an even more ancient era.[13]

Yet even this expansive vision of fire technology is too narrow, because humanity commenced to cook the Earth itself. Fire is a biochemical reaction, not just a physico-chemical one. Entire landscapes, not simply particles of wood and stone, could be fed to it, and undoubtedly were. Instead of bringing objects to the flame, the flame could be brought to a wider world. The hominin hearth expanded over large settings, and eventually the planet.

All organisms shape their habitats. For a savanna herbivore, the geography of burned patches is the geography of food, and this is equally true for the omnivores and carnivores who follow them. By grazing preferentially on greened-up patches, grazers, from African bovids to North American bison, reduce the fuels available for fire and thus help regulate the patterns of burning. But imagine the further power if they could themselves start those fires. They would rule that biota. They would have the power not simply to crack open a bone here and there or to render edible tough biomass but to engineer the ecosystem. The true power of controlled flame was less to drive than to draw, that is, not so much the ability to force prey to flee or rush into hunting sites as to entice them to regular feeding grounds. Such fires are not tamed so much as captured, like training a wild animal to do tricks; control is always tentative, and the nominally tamed creature can go feral. Yet in a landscape predisposed by nature to burn, the ability to strike at will and preemptively conveyed enormous power.[14]

As with simple physical pyrotechnologies, this biological one expanded. It became the basis for fire hunting and fire foraging and fire fishing (the lights would draw fish to where they could be speared). It became the foundation for agriculture outside of floodplains, with the cycle of slash-and-burn emulating the cycle of postburn recovery. It could promote edible tubers and seeds as well as cook them; the hominin diet would be twice fired. Moreover, a pyric synergy came into play such that fire-improved hunting, foraging, farming, and herding tools could interact with fire-catalyzed landscapes. Eventually the fire would be used not only to make machines but to power them. If, as Cicero observed, humans had made a second nature out of the first they inherited, they did so with fire. They had their prehensile hands on the levers that controlled the dynamics of the landscapes they inhabited.

Such reach was beyond the grasp of *Homo erectus*. Fire as Archimedean lever was capable of moving the planet, but only if head and hand could place it properly, and such capabilities were likely beyond the erectines. Still, the torch they lit would never be extinguished. Earth, a uniquely fire planet, had found a keystone species to manage its flames, and the Pleistocene's informing ice had met its most implacable challenger. The erectines were Earth's first Prometheans.

Fire in the Mind

Whether or not fire helped increase cranial size, it helped fill that cranium with lore and symbols. Fire became a standard theme of mythology and folklore, with fire gods as emblems of the inventive, the defiant, and the capricious. Controlled flame came to stand as an enduring emblem of enlightenment, learning, and invention for a long cavalcade of major thinkers who have granted to fire a powerful role in what makes us human, though few did more than contemplate the allusion (an unusual number were mystics). The pre-Socratic natural philosopher Herakleitos made fire the fundamental operation and symbol for nature. That seemed plausible in a time when fire was prominent in quotidian life, when open flame heated and lit up houses, cooked food, forged metals and hardened pottery, refreshed pastures and readied arable fields. But those flames, even if abstracted into symbols, have continued into modern times and in those disciplines most relevant for Pleistocene studies.

Carl Sauer, a geographer, put fire close to the origins of humanity's environmental powers, what enabled early humans to break "the limitations of environment that had previously confined him" and start "a new way of living." Loren Eiseley, a naturalist, concluded that fire was "the magic that opened the way for the supremacy of *Homo sapiens*" and considered humanity itself "a flame." Pierre Teilhard de Chardin, a paleontologist, likened the origins of thought, that is, consciousness, to a flame "that bursts forth at a strictly localized point," an echo, as Landau observes, "of the divine spark of the doctrine of special creation." Anthropologist Lévi-Strauss accepted fire as the chasm between culture and creatures, parsed the world into the raw and the cooked, and declared that "through [fire] and by means of it, the human state can be defined with all its attributes." The British structuralist Edmund Leach then so rose into abstraction that he declared that people "do not have to cook their food, they do so for symbolic reasons to show that they are men and not beasts." So far had industrial humanity come from its roots and so fatuous had intellectual discourse become.[15]

Even as a symbol intellectuals continue to sense—know in their phylogenetic bones—that fire remains at the nuclear core of what it means to be human. Yet by examining only the char and ash, or the icon and the allegory, they just haven't worked through the how and why. They can't find fire in the archaeological record. They can only see its parts and its legacy. For the rest they must rely on a spark of imagination. But that, too, is part of what it means to be human, for as Plutarch observed, the mind is not a vessel to be filled but a fire to be kindled. Not merely a master technology, fire has become a master metaphor.

The Orthogenesis of an Idea

Evolutionary biologists object to the notion that a directed goal—teleology, for Aristotle; orthogenesis, for latter-day natural philosophers such as Henry Fairfield Osborn—might exist. Rather, natural evolution is open-ended, spreading over history like the scouts of ant colonies searching out food who only afterward leave a distinctive trail for others to follow. Such open-endedness is not the case with the narrative that researchers construct out of those wanderings, which must align with a goal or ending theme. Moving from science to literature, phylogeny will thus slide into orthogenesis.

Evolution Evolves

The Darwinian model of evolution by natural selection provided an alternative creation story. It was one in which humanity arose, developed, and reached its present status from natural material and through natural means. Change occurred through the existence of "fortuitous" variability in organisms, and those which were better suited to the environment of the time were preferentially chosen to survive. There was no special creation, and no foreordained destiny. There was only a continuous jostling between species and settings. Humanity was the outcome of happenstance: usefulness selecting among chance.

That was the pure version implicit in *On the Origin of Species* (1859). Darwin made the argument explicit in *The Descent of Man*, published in 1871. Between those books other commentators eagerly piled in: T. H. Huxley, with *Man's Place in Nature* (1863), Charles Lyell, with *Geological Evidences of the Antiquity of Man* (1863), and Ernst Haeckel, with *Generelle Morphologie der Organismen* (1866). Meanwhile, in a cave overlooking the Neander Valley, the first fossil hominin, what became the eponymous Neanderthal, was found in 1857. A similar skull, first uncovered in 1848 and forgotten, was rediscovered in Gibraltar in 1863. Over the next few decades partisans scoured caves throughout Europe and revealed a horde of prehistoric hominins:

Neanderthals; Cro-Magnons; denizens of the "ages" of the hippo-
potamus, of the cave bear, of the mammoth, of the reindeer. Stone
tools, reworked bones and antlers, rock paintings, hearths—all testified to peoples who flourished prior not only to contemporary times
but to any recorded in the Bible or the most comprehensive works of
the ancients.

More relict bones appeared from more places. They came from
caves in Belgium, from eastern Europe, from the Levant—Iraq, Pales-
tine, and Israel. All seemed to display traits typical of the Neanderthals. Surprisingly, some came from China and Java; these seemed
more primitive, erect but with crania characteristics intermediary to
apes and humans. Places that lacked such relics might invent them, as
Britain did with Piltdown Man or America with Nebraska Man. Each
find acquired a new name, and typically a new genus—*Zinjanthropus*,
Hersperopithecus, and *Titanhomo,* to name a few. Each declared itself
the missing link in the great chain of hominin evolution. The upshot,
however, was mounds of links in museums without much of a chain
to join them. By the 1930s the proliferating collections had become a
zoo, desperate for a new Linnaeus to sort them out.

More data would seem to improve understanding, but these seemed
to scramble it. Why? More than a shift of paradigms, the understanding of what constituted an explanation was itself undergoing a paradigm shift. The challenge came not from fossil hominins as anomalies
but from the general character of the theoretical construction into
which the hominin fossils were placed. In its original formulation,
Darwin's theory had relied on chance and natural selection to evolve
new species from ancestral ones; the process was incremental and universal, a declension of forms from a common source. Over the years
every part of that founding conception fell apart, as fossils poured in
from elsewhere, national or ethnic narratives replaced global ones, and
Darwinism, at least in its originating formulation, began itself to
descend, iteration by iteration, into other species of idea and "law."

The fact is, Darwin's theory had very different receptions around
the world. It had to adapt to existing frameworks of thought, not only
to biological theories but to the social settings that translated data

from natural science into cultural meaning. "Darwin" thus meant very different things. In Britain it resonated best, since the view of the world it promoted emerged from the experiences of that society—and which is why the codiscoverers of the idea (Darwin and A. R. Wallace) had both grown up in a common culture. France refracted Darwin through the prism of its own chauvinistic heritage, which translated Darwinism into a modern restatement of the "transformism" that French intellectuals had long pioneered. Germany bonded Darwinian evolution to its extensive laboratory traditions in embryology, of which Ernst Haeckel's biogenetic law (ontogeny recapitulates phylogeny) is an apt expression. Russia received Darwin as it was experiencing a social revolution, the emancipation of the serfs, and an economic liberalization; it saw the evolutionary record as one that supported notions of progress. America quickly accepted an evolutionism shorn of Darwinism. The *Origin* arrived on the eve of a civil war and then hybridized with old conceptions of providential history and progress.[16]

Overall there was little appreciation that history, like Darwinism, might rely on the fortuitous and the purely utilitarian. Rather, each new hominin fossil marked a milestone on the road to perfection. This, however, was a narrative that could be remade by each society and could be rewritten to fit each's own national or ethnic creation story. Evolution could splinter apart as much as rise upward, such that some parts of humanity could evolve faster than others. Thus, in Haeckel's hands, human evolution became polygenetic. To less conscientious minds, it substantiated a nominally scientific racism. But in an era when European imperialism had reached flood tide, most thinkers were inclined to imagine competing European nationalisms and their overseas expressions as the master narratives. Africa was a recipient of colonizers, not a source for them. There seemed no reason to place humanity's origins in what had been labeled a dark continent.

Scientific challenges also mounted. Recall how Kelvin (incorrectly) questioned the reservoirs of geologic time that Darwinian evolution demanded. Even Darwin, equipped with inadequate genetics, struggled to retain the original theory, and modified it in successive editions of the *Origin*. By 1900, when Mendelian genetics was rediscovered, Darwin's particular depiction of evolution had receded to mostly British practitioners and yielded to popular conceptions that equated evolution with progress. Evolution as a narrative of design in history thrived; evolution as a game of chance in which the dice was rolled on an ever-changing table had few bettors.

Then the parts, as scattered as the fossil bones of erectines, began to come together. A generation of lumpers replaced one of splitters. Genetics and gene pools, a revived paleontology and a more robust conception of systematics, an embryology shorn of evolutionary allusions, a reaffirmation of natural selection—all converged into what became a new understanding. The idea found its first full expression when Julian Huxley, a nephew of T. H., published a book titled *Evolution: The Modern Synthesis* in 1942. Interestingly, many of Huxley's formative field experiences had unfolded in Africa, beginning in regions not far from the Rift Valley and at a time when the Leakeys were engaged in their excavations at Olduvai Gorge. Others promoted the emerging idea as neo-Darwinism or, the most widely used expression, lifted from Huxley, the modern synthesis. Classical Darwinism, like a heritage building with its guts ripped out and replaced by modern wiring and furnishings, revived, and with it Darwin's hunch that hominin evolution might point to Africa. By the postwar era imperialism was fast receding before a wave of decolonization, wars had tarred most nationalist narratives, and the world seemed willing to establish a United Nations of narrative.

The study of early hominins followed these trends. There was more of everything, and it all began to converge: more and diverse fossils; reams of data from mitochondrial DNA; a much denser and more finely textured chronology of Pleistocene climates and landscapes; the renewed conviction of natural selection (a collateral benefit of neo-Darwinism); a newly liberated Africa eager to celebrate its heritage; and an inclination toward universal history. All these comingled into a consensus that humanity had emerged in Africa and had thence spread over the Earth, and that hominins had done so not once but repeatedly. The geographic and genomic narratives created a kind of modern synthesis for paleoanthropology, what became known in shorthand as the Out of Africa theory.

It appeared to some that the study of hominin history had finally purged itself of metaphysics and had emerged from the refiner's fire as a genuine science. The upward progression of the discipline seemingly echoed the phylogenetic record of hominin evolution. If so, they ignored the deeper context of their inquiry, the one that could connect the evolved big brain to the world which inhabited that brain. The chronicle of evolution and migration had to become a narrative, for which they needed the kind of meanings that only cultural scholarship could convey, and to which the expression Out of Africa might itself stand as synecdoche.

Out of Africa

Even in Latin it is one of the commonplace phrases from recent times, or, granted its recitation in a dead language, perhaps one of the era's most enduring clichés. *Ex Africa semper aliquid novi*: "There is always something new coming out of Africa." By the time it served as shorthand for the reigning narrative of hominin origins, it came packed with contemporary judgments: that novelty was good; that humanity had evolved and migrated; that Africa, lagging so far behind the rest of the world by standard measures of importance, should be prized. The richer value of that phrase, however, may be its own evolution.

The expression, as Harvey Feinberg and Joseph Solodow have traced it, originated in Greece as a proverb "no later than the fourth century BC." The playwright Anaxilas recites it in a play mocking Plato. Aristotle incorporated it into both his *Historia Animalium* and his *De Generatione Animalium*. In all these instances the expression refers to hybrid animals, unknown elsewhere, along with a sense of an oddity that hedges into the monstrous. The phrase then sank into the general cultural background, the bovids of the ancients' literary savanna.[17]

It reappears, with mutations, in the second century. The paroemiographer Zenobius included it among his compilation of Greek proverbs, of which several variants achieved circulation. "Newness" by now acquired a connotative meaning of "evil." More famously, Pliny the Elder, acknowledging his debt to Aristotle, inserted yet another version into his *Natural History*. In the process of translation, like a DNA transcription error, the Greek verb *ferein*, which means to "bring forth" or "give birth to," became the Latin *adferre*, which means "to produce," but without the sense of birthing. Novelty, if not evil, comes out of Africa in more than its wildlife.[18]

At this point, at least as recorded in recovered texts, the stock splits. The Zenobian clade dies out. The Pliny clade disappears into the sorry loss of learning after the collapse of the Western empire. It then was preserved through Arabic or Byzantine archives until it was recovered in the Renaissance. It emerged with special glitter out of the era's textual caches when Erasmus included it in his *Adagia*, which appeared in a swarm of editions—some 250 in all—beginning with the *Collectanea Adagiorum* of 1500. The epitome of the Renaissance humanist, Erasmus cites Pliny as his source but also tracks the phrase's etymology to Aristotle and Anaxilas. Interestingly, his resuscitation again altered the actual wording, this time bequeathing *Ex Africa semper aliquid novi*—literally, "out of Africa always something(s) new." *Africa* replaced the (by now) more local *Libya* and the troubling verb was dropped. It is suggested that these changes may in fact reflect another error in transcription, the "hurried reading" of a marginal notation.[19]

"There can be little doubt," note Feinberg and Solodow, "that from Erasmus the phrase acquired such renown as it enjoys, both in the Renaissance and in modern times." If the fourth-century Greeks were the habilines out of Africa, Erasmus paralleled the erectines. His phrase replaced the original and spread widely in its subspecies variants. From Erasmus, Rabelais adopted the expression in *Gargantua and Pantagruel* (1536) as did John Donne in his "Satyre IV." Then matters paused as the classics fell from favor relative to the *novum organum* of the emerging sciences. In the eighteenth century Alexander Pope echoed the expression in a passing text, curiously retaining its sinister connotations. Otherwise, the phrase retreated to a South African refugium. In 1877 the South African Museum adopted it as a motto.

In the 1920s the phrase could still be found circulating amid the educated elite of Cape Colony. But this may be less a sign of robustness than of isolation. As its larger environment—classical education—collapsed before the chill of modern science, its use survived in protected enclaves where classical allusions to Africa might still be welcomed.[20]

The modern variant appeared in 1938 through the English translation of Isak Dinesen's *Den Afrikanske Farm*. Somehow, in moving from Danish to English, the old proverb was resurrected once again and attached prominently to the title, which now read *Out of Africa*. Africa became prominent in the postwar era, as it shed its colonies and became an object of geopolitical interest to cold war rivals. Other authors picked up the phrase from Dinesen, each leading to others, until the process reached a climax in the title to a major 1985 movie based on her book. By then a new linguistic radiation was under way that propagated an erstwhile expression from ancient Greece throughout popular culture.[21]

So when paleoanthropologists sought to label their narrative of human origins, they could latch onto a proverb that floated through the cultural ether. "Out of Africa" became the title of the 1996 Louis S. B. Leakey Symposium held at Stanford University that effectively confirmed the contemporary understanding of hominin history. If the phrase could not by itself convey the larger narrative, the story of how it came to be applied might do better, and could successfully recapitulate the larger theme, which was not the emergence in and migration out of Africa by hominins but the disjunctive and fragmentary record of ideas, and how they may be recovered and reinterpreted. While modern research excavated the fossil records of biological evolution, it was a long tradition of cultural evolution that explained what they meant.

From Tangled Bank to Braided Narrative

It was how that transfiguration had happened that perhaps holds the most interest. In concluding the *Origin of Species* Darwin imagined "a tangled bank" overflowing with living forms yet organized by discernible laws, and while full of "grandeur," a scene that did not result from a preformed pattern. Yet as Ernst Cassirer has argued, "Man cannot escape from his own achievement." Darwin's tangled bank has been replaced by a "tangled web of human experience" that weaves together language, myth, art, religion, and all the other strands of humanity's "symbolic net." That peculiar capacity of human thought remade Darwin's tangled bank into a shelf of braided narratives in which the entwining of genomic and geographic data had to play out over a cultural landscape: that was where, to continue the analogy, the selection would take place. The revival of neo-Darwinian concepts, however, too often brought with it a neo-Darwinian scientism that failed to apply to its own informing conceits the perspective it demanded of others. In particular, it made Darwinian evolution an act of special creation.[22]

It was a simplistic narrative that assumed that ideas could be discovered out of data the way bones could be found in sandstone or tuff, and it viewed the progress of biological science (and archaeology) in a way partisans scorned when others applied it to their own fields. They did not appreciate the extent to which their explanatory ideas, even the theory of organic evolution, had a long history, and that, like *Equus caballus* within the equids or *Homo sapiens* within the hominins, the idea was not the intended end product toward which all research had trended but the selected survivor of ancient stock, a product of happenstance, historical contingency, and usefulness. Disciplinary histories tended to be teleological, as narrative must be; the history of the idea of evolution was thus orthogenic in ways the theory's advocates denounced when applied to nature.

Darwinian evolution was less a special creation, the spark of a divine insight, than it was the rough, imperfect, best adapted, useful, and cantankerous outcome of a tedious and often errant chronicle of observations and imaginings. It was a powerful idea, and once discovered, destined (so it seemed to many) to ramify across whole continents of learning. It offered a promised consilience, which could seem the apex to which all prior study had tended. But such apparent inevitability was an inherent construct of narrative, and just as an organism's traits are not intrinsically better or worse but better or more poorly adapted to its setting, so it is with ideas. The evolutionary paradigm achieved much of its power and reach because it tapped into very old traditions of thought. Far from being a radical innovation without precedent, Darwinian evolution had itself evolved by fits and starts out of one of the hoariest concepts in Western civilization, the Great Chain of Being.

Chapter 6

Missing Links

See through this air, this ocean, and this earth,
All matter quick, and bursting into birth.
Above, how high, progressive life may go!
Around how wide! how deep extend below!
Vast Chain of Being!

—Alexander Pope, *An Essay on Man* (1732–34)

"MISSING LINK." No other phrase so encapsulates the search for and interpretation of fossil hominins—or, for that matter, the vast menagerie of once flourishing and now vanished creatures from the Pleistocene as this hoary term. The expression is itself somewhere between a snippet and a synecdoche—a shorthand, even—for a majestic intellectual project that dates to the foundations of Western thought, when mythology moved into philosophy before transmigrating into successions of sciences. That conception is the Great Chain of Being, and like its subject consists of a plentitude of ideas, discoveries, connotations, personalities, and events, all linked by a continuous gradation that has expressed an invariant if inchoate yearning for order. The Chain of Being was more than a program of science: it was, in a single expression, a full-bandwidth summary of all that could appeal to nearly every form of inquiry. In the words of E. M. W. Tillyard, the chain was a "metaphor that served to express the unimaginable plentitude of God's creation, its unfaltering order, and its ultimate unity."[1]

For centuries the chain as a body of empirical data and a conceptual ordering of ideas built up, link by link, with new links supplied by reimaginings and discoveries and expansions. It achieved a first-order form in ancient Athens. It reached a double apex in the eighteenth and nineteenth centuries. In the first, it completed its inherited task—providing a cosmogenic order to the Earth's cornucopia of objects, from minerals to the Moon. In the second, as still more discoveries threatened to shatter that arrangement by proliferating candidates for links faster than they could be reforged, the order assumed another dimension and became historical. It was during this period that the Chain of Being furnished both a conceptual matrix for understanding where fossil hominin discoveries might fit and an agenda for discovering them. No conceptual apparatus so shaped the search, interpretation, and public reception of fossil hominins more than the Chain of Being. Missing links became an *Objekt* in the Cassirerean sense. They were at once thing, history, and meaning.

Then the intellectual scaffolding collapsed—partly from its own weight, and partly by challenges from the larger culture, which chose to see the world through different philosophical prisms and to express that understanding through new matrices of science and tropes of literature and streams of evidence. The Great Chain that had once bound the lowliest flake of mud to the most distant stars, and that had granted early hominins, both found and fancied, a distinct place in the order of things, disintegrated.

Yet the extinction of the Great Chain as an organizing conceit was only apparent. The concept is itself a missing link in understanding the study of early hominins and the Pleistocene, for the persistence of its underlying forms was bonded to an enduring moral sense. Revealingly, the Chain was first given explicit expression by Plato in the *Timaeus*. Later, explaining his efforts to systematize Platonism, Gottfried Leibniz asserted that Plato's ethics and metaphysics went together "like mathematics and physics." So, too, it has been with that other Platonic legacy, the Chain of Being. It contained both the material and the moral. And like a Platonic type, if much battered and misshapen by use, it stands for the persistence of certain kinds of questions, particular sorts of answers, and a ceaseless discourse on distinctive controversies.[2]

The Chain of Being as a formalized intellectual project achieved its apex just as the Pleistocene was developing as a concept. The instrument of transition for both was the discovery of a new world— the dominion of geologic time. Over the next 150 years the known age of the Earth expanded a millionfold and lost worlds of the past were found to have overflowed with species now gone. The chain stretched, aged, and eventually succumbed to senescence. The Pleistocene, too, grew older and acquired new traits. Two centuries after they had joined, the two ideas parted.

Or so it seemed. As the Pleistocene was having its temporal boundaries redefined at the onset of the twenty-first century, the Chain was being treated as a relic shorn of its provenance, a fossil dug up from the past and stored in a cabinet of curiosities. Yet the deeper instincts behind the Chain of Being have persisted, and they have found new ways to bond the material world with the moral. It is a fable deftly illustrated by the intellectual saga of the mastodonts, whose excavated bones first demonstrated the reality of extinction, forced the chain to become historical, and became themselves an icon of the Ice Age.

Forging the Chain

What became known as the Great Chain of Being was for Western civilization, across perhaps two millennia, the "most widely familiar general *scheme* of things, of the constitutive pattern of the universe." It was, as A. O. Lovejoy exhaustively explicated, a popular phrase, a metaphor, a research program, and a concept both metaphysical and moral. It explained the why, the what, and the how of the world. It characterized humanity's identity by specifying its place within that larger universe. Its chain of thought linked popular sensibilities, natural science, theology, philosophy, and literary imagination.[3]

Its life cycle, as E. M. W. Tillyard has distilled that history, began "with Plato's *Timaeus*, was developed by Aristotle, was adopted by the Alexandrian Jews (there are signs of it in the philosopher Philo's work), was spread by the Neo-Platonists, and from the Middle Ages till the eighteenth century was one of those accepted commonplaces, more often hinted at or taken for granted than set forth." From mythology early philosophy abstracted, like a being emerging from the chaos, a principle of self-sufficing perfection, or what later simplified into a principle of sufficient reason. For the religious, this meant the world existed as a manifestation of God, and thus displayed the deity's properties. For the secular, it meant that the world existed in and for itself, but again, as a fully realized expression of nature. From Plato the principle became an Idea and received a Form. The idea was the principle of plentitude, which stated that the world is a plenum that holds everything possible. The form was the image of a Chain of Being in which all the pieces of the plenum are joined in strict and comprehensive order. From Aristotle came the impetus to organize the parts into a grand classificatory schema, a *scala naturae*.[4]

The Chain embodied two other principles: continuity and gradation. It had no gaps, no lapses, no voids. And the links that bonded one part to another had an order expressed as a sequence and hierarchy, such that no link was interchangeable with another. Every link held the properties of all those below (or before) it, to which it added a new trait of its own. The upshot was a comprehensive schema that included everything and gave to everything its own place, as a breakdown in any part would be catastrophic to the whole. Unsurprisingly, it emulated the social order that existed, or that intellectuals wished to have exist.

"The result," Lovejoy argued,

was the conception of the plan and structure of the world which, through the Middle Ages and down to the late eighteenth century, many philosophers, most men of science, and, indeed, most educated men, were to accept without question—the conception of the universe as a "Great Chain of Being," composed of an immense, or—by the strict but seldom rigorously applied logic of the principle of continuity—of an infinite, number of links ranging in hierarchical order from the meagerest kind of existents, which barely escape non-existence, through "every possible" grade up to the *ens perfectissimum*, . . . every one of them differing from that immediately above and that immediately below it by the "least possible" degree of difference.

By virtue of its reason (or soul), humanity occupied the link between the brutes and the celestials. By the time of the Renaissance, William Shakespeare could dramatize the human condition as the rung between Ariel and Caliban.[5]

The Great Chain answered the grand questions. In the principle of sufficient reason, it found a "why" to the world. In the principle of plentitude, it discovered the "what." In the principles of gradation and continuity, it elaborated on the "hows." It was comprehensive in its scope, rigorous in its details, elastic in its ability to absorb new evidence. No other explanatory system came close to offering an alternative. Everything was, in Pope's inimical couplet, but a part of one stupendous whole: "Whose body Nature is, and God the soul." It all pivoted around humanity. "The proper study of Mankind is man."[6]

All in all, it was an extraordinary conception, welded into a hard system—and idealized, perhaps—by neo-Platonists, and eventually bequeathed to the Renaissance. From humanists it spilled into literature, which found ample outlets for such an organizing conceit. Shakespeare's plays are full of references to the Chain (or sometimes "scale") of Being, and it serves as the theme of the celebrated set piece by Ulysses in *Troilus and Cressida*.

The valence between theology and philosophy, which was present from the start, was restated in terms more at ease in the Renaissance. Its *summa* was Raymond de Sebonde's *Natural Theology* (1550). Here the vast chain rose from inanimate minerals, each ranked by its unique "virtue," through the vegetables, the animals, and the celestials. Humanity "sums up in himself the total faculties of earthly phenomena," as Tillyard explains, and "for this reason he was called the little world or microcosm." Symmetry then demanded that the celestial realms above have the heft and complexity of the material world below, for "[t]here is a progression in the way the elements nourish plants, the fruits of plants beasts, and the flesh of beasts men." All tend toward God, and it took no great leap of imagination—the mind of the era being no less hostile than nature to leaps—to see the Chain of Being as a manifestation of the divine will, and that to know one was to know the other. Natural theology was to theology as natural philosophy was to philosophy.[7]

This seems strange to modern sensibility in which theology and science must quarrel. It made perfect sense to the Renaissance, however, and as modern science distinguished itself from humanism and other scholarships, shedding the books of the ancients for the book of nature, it retained the Chain of Being as a formative principle. Natural philosophy was not yet ready to sever its links to the metaphysical and the moral. Leibniz put the Chain at the core of his system, for the concept seemed to span physics and metaphysics as humanity did the crass and the celestial. But the schema appealed no less to John Locke, who saw "no chasms or gaps" in "all the corporeal world," and who imagined that "the species of creatures should also, by gentle degrees, ascend upwards from us towards his infinite perfection, as we see they gradually descend from us downwards." In such an arrangement Locke perceived "the magnificent harmony of the universe" and a testimony to the "great design and infinite goodness" of its Architect.[8]

For the professional naturalists of the era the Chain offered more than a dazzling metaphor. It proposed a program of research. Assumptions latent in the scheme became hypotheses; ideas implied became injunctions to discover confirmatory evidence. The project became in modern parlance a paradigm. It told normal science what to do: how to gather suitable data; how to organize that collected information;

how to interpret the significance of parts and whole. It provided a system. Their success became, in turn, a measure of the Chain's power, and its weakness.

In such ways what might have remained a somewhat loopy literary trope or a feral metaphor from metaphysics became an organizing principle for emerging sciences. "Purely speculative and traditional though it was," Lovejoy observes, the Chain acted on natural history as "the table of the elements and their atomic weights" did on chemistry. Despite the scaffolding's increasing secularization, its champions did not shirk from placing humanity itself within it—that, after all, had always been the purpose of the project, to explain how humans fit into the scheme of things. So even as intellectuals substituted Nature for God, people still stood between worlds—between matter and spirit, between natural philosophy and humanist scholarship, and between instinct and reason. People were the highest of creatures and the lowest of spirits. As naturalists sought to fill in the nominal gaps between flowers and bees, sandstones and carnivores, they found that humanity plugged the gap between apes and angels, and they realized that there were gradations between peoples as well. To complete the Chain those missing links, too, would have to be found and fitted.

That the eighteenth century should have brought the enterprise to a climax seems both inevitable and surprising. It was an era of codification and system, a middle link itself between the worshipful attitude of the Renaissance toward ancient learning and the modernist's scorn of anything but the new. It filled the Chain, lavished the idea of the Chain with abandon, and perfected it as a catalog of natural history and a principle of moral philosophy.

In no other time, Lovejoy concludes, did "writers of all sorts—men of science and philosophers, poets and popular essayists, deists and orthodox divines" so accept and obsess over the scheme and its implications. It penetrated every field of inquiry. In the *Systema Naturae* of Linnaeus and in Alexander Pope's *Essay on Man*, it found its fullest expressions. Yet in those supreme achievements its architects anticipated one of C. Northcote Parkinson's sardonic laws, that the "perfection of planning is a symptom of decay . . . achieved only by institutions on the point of collapse." Perfection of an edifice— whether made of ideas or stone—comes "when all the important work has been done." For perfection of the structure means finality.[9] The perfection of the Chain revealed in Linnaeus's grand classification was based on "artificial" (that is, arbitrary if useful) traits. The perfection of Pope's *Essay*, coded in heroic couplets, while magnificent was no less artificial, and far removed from everyday language and learning.[10]

The very universality of the Chain, moreover, left it increasingly detached from meaningful discourse. It became a cliché. It was, as Lovejoy put it, "obviously not a generalization derived from experience, nor was it, in truth, easy to reconcile with the known facts of nature." It was neither true science nor sound philosophy, and it came to assume a status as shorthand for the way things are, akin to the expression "Establishment" in recent decades. Skeptics such as Voltaire and Samuel Johnson ridiculed both its precepts and its practice. Its value as a description of humanity's ethical quandry edged into secular platitudes. Its proposed categories of yet to be known beings were either fanciful or had little prospect of discovery, while new facts proliferated that didn't fit. It appealed to literary folk especially, in the way psychoanalysis did in the early twentieth century, and it just as quickly sagged from its own excesses and exegesis. Once a dominant figure, it became a part of cultural background chatter.[11]

From Chain to Tree

Flush with the triumph of Newton, Linnaeus, and the *Encyclopédie*, the Enlightenment had rebuilt the West's understanding of the world along modern lines, which is to say, according to reason and the *novum organum* of empiricism and mathematics. Yet in curious ways it was a static world, fixed in its mathematical rigor and clockwork mechanisms, awaiting only the fuller enlightenment of higher math and more clever experimentation. It thus tended toward the elaboration and ornamentation of the known, indulging in the Baroque and then the Rococo, before asserting the restored purity of a neoclassicism that reflected the assumed fixity of the natural order and the society that presumed to emulate it. It turned the misadventures of Robinson Crusoe into a cautionary tale about wanderlust and ambitions beyond one's station. It satirized explorers as so many Gullivers.

But a wave train of revolutions swept over the landscape. Political—the American, the French, those of emerging nationalisms. Economic—the industrialization of power and commerce. Social—the breakdown of hierarchy, established institutions, and aristocracy. Geographic—a second great wave of discovery that bonded empire and Enlightenment and burst beyond the limits of what might be known and ruled. Intellectual—the transfer of knowledge to a literate public, the efflorescence of a natural history robust enough to challenge natural philosophy, and the belief that humanity's surroundings were not imperfect facsimiles of an ideal order, immutable from the Creation, but something that could grow into perfection. The contemporary world need not, at best, re-create the world of the ancients, but could become something new. It could grow. It could know progress.

The Chain of Being had built upon the existence of many species and assumed many more that were not yet known but were needed to fill out the roster of links. It did not, however, allow for links that had once existed and then disappeared. The assumption was that they would be found in other lands not yet explored. Instead it appeared that they came from other times. In fact, there appeared to be whole worlds that had flourished and died out in the past. How to reconcile those lost worlds with the imperative expressed in a Chain of Being? Some thinkers assumed the nominally extinct species would ultimately be found in the yet unexplored lands of the Earth. Thomas Jefferson, for example, instructed Meriwether Lewis to look for mastodons in the Rockies. Others thought that, on the model of the Noachian flood, the world had been created and destroyed over and again by massive convulsions. Georges Cuvier, a founder of comparative anatomy and paleontology and the first to reconstruct a mammoth, thought the Earth had experienced half a dozen such catastrophisms. Each epoch had created a new chain. As more discoveries poured in, however, it became possible to smooth out the transitions in the stratigraphic record, and, in time, the sequence of species lost and found came to resemble itself a chain, albeit one through history.

Behind all this stood a new era of geographic exploration that launched Enlightenment expeditions across Earth's continents. In this second great age of discovery, as William Goetzmann has termed it, the returned plunder included plants, insects, rocks, exotic marsupials, birds, rivers and mountains, and of course a thick volume of previously unknown peoples. At first, existing systems could accommodate the bullion, and then a kind of inflation overwhelmed the economy of knowledge, and finally the codifying systems became bloated and deformed and collapsed under their own weight.[12]

The most prominent naturalist of the age, Linnaeus, devised a method to organize the flow, at first a trickle, and then a flood. He published the first edition of his *Systema Naturae* in 1735, the year La Condamine commenced his famous expedition to Ecuador to measure an arc of the meridian; the thirteenth, and last, in 1769, the year the world launched a fleet of expeditions to measure the transit of Venus, among them the first of Captain James Cook's circumnavigations. The 1735 edition had eleven dense pages. The last edition had three thousand. The tenth, published in 1758, is credited with establishing the modern binomial system for classifying animals by genus and species. The outcome was a system for naming everything— Linnaeus was dubbed a "second Adam"—but not a system for containing everything so named. Even as a trope, the Chain of Being could not absorb them all. And then the floodgates of discovery opened.

Such revolutions in formal learning and felt experience needed new modes to express their understanding, beyond sonorous Ciceronian periods and the rigid structure of heroic couplets. It needed something looser, longer, and above all historical, which could not only link events, data, ideas, and context through time, but in which history could itself serve as an informing principle. The age craved creation stories in which the logic and moral order were manifest in and through the unfolding of the story.

The historical mode—in its literary manifestation, narrative—infected every field of study. Natural philosophy added a nebular hypothesis to account for the origin of the invariant mechanics of the Newtonian solar system, and it created a new chronometer through the second law of thermodynamics. Natural history became overtly historical. It reformed biology along temporal lines, invented geology, and pondered the age of the Earth and the evolution of life. Philosophy contemplated changes over time and then did what philosophers had always done—identified fundamental principles and systematized its conclusions. From Georg Wilhelm Friedrich Hegel to Karl Marx to Herbert Spencer, the century turned to history for explanations: to know something's origins and its mechanisms of change through time was to explain it. So, too, literature moved beyond essays and epistles and into novels, lush with subplots and complicated story lines. Formal history followed that lead; this was the era of Romantic history, of nationalist creation stories, of grand narrative. Moreover, to know the past was to predict the future; from something's origins you could trace its progress to come.[13]

The Great Chain of Being became, in Lovejoy's phrase, "temporalized," or in an expression of metaphoric metempsychosis, it became organic. Its links were stages of development. What had served as an all-purpose metaphor for the workings of the universe, a clock, morphed into that of a tree. The principle was universal, such that becoming characterized the smallest entity no less than the whole. It was fractal, in that the same principle expressed itself at every scale. And it embodied design, in that it was no more arbitrary or fortuitous than the maturation of an acorn into an oak. According to Pope:

Where, one step broken, the great scale's destroy'd
From Nature's chain whatever link you strike,
Tenth or ten thousandth, breaks the chain alike.[14]

Ernst Haeckel: Archetype of an Era

When in 1898 Ernst Haeckel looked back on the prior half century
and declared it "the Age of Darwin," he did not need to state that he
had done much to make it so. With the idea of evolution he had fused
biology to comparative anatomy, paleontology, and embryology, created an iconic image to replace the Chain, and made explicit the
meaning of the new organon for understanding humanity's place in
nature. "[A]nthropogeny," which he regarded as the "most important
conclusion of the theory of descent," exceeded "all the other truths in
evolution."[15]

The process originated in 1866 when he announced the biogenetic law, according to which the development of the individual re-
created the development of the larger group to which it belonged.
Stated in formal terms—which Haeckel invented—ontogeny recapitulates phylogeny; that is, the embryological development of the
fetus passes through the evolutionary stages by which the species
had emerged. Each individual thus carries the larger narrative within;
they both share a sense of patterned historical development. It is not
that Darwinian evolution is the model for embryology; rather, embryology is a model for evolution, as life moves from the simple to the
complex, and from amoebas to people. Apart from its specific decla-
rations, the biogenetic law confirmed the universality of history as
an organizing and explanatory medium. The fundamentals of the
world did not operate like the gears of a clock but the branching of
a tree.

How this happened Haeckel explained in *The History of Creation*
(1876), in which he elaborated on humanity's place within the Tree of
Life by mixing metaphors and identifying the Chain's links with
stages of evolution. He calculated that twenty-two links in all spanned

from formless protoplasm to humanity. The twenty-first stage in the "Chain of Animal Ancestors of Man" was hypothetical but necessary: an apelike man, which Haeckel labeled *Pithecanthropus alalus* ("speechless ape-man"). Even in this vitalized restatement, the logic of evolution—of historical plentitude—required intermediate beings. And while Haeckel admitted that "we as yet know of no fossil remains" of this creature, it took "but a slight stretch of the imagination to conceive" of such "an intermediate form." It was, in truth, a logical necessity. Its "certain proof" could be reached "by an inquiring mind" alone, since the logic of creation demanded it. But full conviction required fossil evidence, which Haeckel thought would come from southern Asia.[16]

In 1879 Haeckel completed his hat trick by giving his understanding a formidable image. A figure in his *Evolution of Man* replaced the Chain of Being with a Tree of Life that depicts the phylogeny of everything. The "temporalized" Chain he redrew into a vast tree, which grew from roots to branches with an upthrusting trunk that ended with an apical climax. The image invested an intellectual life force into a discipline dedicated to the study of life; the subject defined its own interpretation. Moreover, it complemented the message of the biogenetic law. The tree grew as an embryo did, not as an open-ended sprawl but according to encoded principles that governed development. Not least, the image tapped a folk sense as old as the Chain, for the tree was to many cultures a symbol of life. Nordic lore, then enjoying a certain fashion among intellectuals, had its Yggdrasil, the World Tree, and Haeckel in effect redrew it for the scientific literature. Like a tree itself, German *Naturphilosophie* thus rose from folk roots into laboratory science.

There was nothing accidental about development. It was a kind of secular providence embedded into the behavior of the world. Without it, progress was impossible, and so was science. Haeckel agreed with Herbert Spencer that "progressive heredity is the indispensable factor in every true monistic theory of Evolution," and that in its absence one would be better advised "to accept a mysterious creation of all the various species as described in the Mosaic account." Three times, Haeckel affirms, he had visited Darwin at Down, and "on each occasion we discussed this fundamental question in complete harmony." The issue was not just that evolution needed laws to work by but that science needed such a lawful universe if it was to explicate those workings. Otherwise, the Chain would have gaps and the tree would be marred by lost limbs or a gashed trunk.[17]

Haeckel's influence, especially on the continent, was powerful, and German learning, embedded in its universities, attracted students from around the world much as American universities have in recent decades. There was nothing modest about his ambitions, as he searched for universal principles and forms and proposed to merge philosophy, art, and science, which he did in both encyclopedic tomes and popular essays. In 1868 he published *The History of Creation*; in 1895–99, *The Riddle of the Universe*; and framing both, various editions of *The Evolution of Man*. Like Linnaeus he was a "compleat naturalist," who traveled to the Canaries and the Indies and worked over data from the *Challenger* expedition; brought system to the metastasizing collections of geographic discovery and laboratory; and captured the zeitgeist of his age.

His works became more than textbook digests of existing knowledge. They were a manifesto to complete the phylogeny of the tree, to substantiate its mechanisms, and to trace its branches, especially that final chain from ape to human. Thus, as the century ended, he published in 1898 a treatise revealingly titled *The Last Link*. The lost link was the last link needed to complete the living chain. As the nineteenth century concluded, biologists pursued that missing link with the dogged determination and zeal with which particle physicists at the end of the twentieth century sought the Higgs boson—or with which naturalists at the end of the eighteenth century had scoured the globe to fill out the links in the Chain of Being.

Links Found, Lost, Invented, and Written

The Chain (or tree) did what scientific paradigms are expected to do. It identified what was missing and what it should look like, and thereby animated efforts to search out the fugitive puzzle pieces. This was particularly useful because those missing links were not discovered in sequence but serendipitously. Like the periodic table, the Chain gave order to what might otherwise remain a cabinet of curiosities or a scatter diagram.

But while the Chain inspired some researchers to find links, it tempted others to imagine them, or still others to simply invent them outright. Later, after the Chain and tree had lost their scientific value and their descent from a creator, they continued to hold their worth as a moral template for stewardship over creation. In the eighteenth century it was unthinkable that links, once forged, might disappear. In the late twentieth it became unacceptable (and to the extent that humanity was responsible, shameful) that links, any of them, should disappear, particularly when human malfeasance was the cause of their loss.

Links Found

Consider two celebrated discoveries as examples of links found. Each was fundamental to an understanding of hominin history and the way the Chain applied to its study. In one discovery an early hominin struggled to find its place in theory. In the other, it was found precisely because its place was already known.

The first example opened the complex saga by which research met the Neanderthals. In 1856 hominin remains were found by laborers in a cave overlooking the Neander Valley near Düsseldorf. The bones were just bones, they thought, and nearly all were tossed aside as rubble before a local teacher, J. K. von Fuhlrott, collected a skullcap and limbs. The bones had no provenance; they came from a quarried cave, and they were tainted by evidence of severe arthritis. Von Fuhlrott presented them to Hermann Schaaffhausen, a professor of anatomy at the University of Bonn, who announced them to the world at a local meeting of the Lower Rhine Medical and Natural History Society in February 1857.[18]

And so it went, as discussion passed through the stages of routine science. The data had a theory to explicate it after Darwin published the *Origin*, and both Charles Lyell and Thomas Huxley acquired casts of the bones and weighed in with commentaries about their meaning. The bones now had a context they lacked in the cave, as they resided in an intellectual setting. Discussion spiraled around the character of the bones and how they might be sited (or not) within an evolutionary phylogeny. All those concerned with the question of human evolution mulled over the bones. All pondered where to place the bones in the *scala naturae*. Authorities like Rudolf Virchow, accenting the acute arthritis, considered them a regressive modern human, while Marcellin Boule, considering the generally archaic features, regarded them as belonging to a descendant of the apes but not an antecedent to humans. As more, similar fossils were discovered, the debate continued in the singsong, give-and-take of normal science. The Neanderthals, described scientifically as *Homo neanderthalensis,* did not unequivocally fit into a recognizable slot in the evolving Chain. It was too far from the apes and too close to humans to qualify as the missing link between them. The dialectic of data and theory continued—and continues to the present day.[19]

The second discovery followed a very different scenario. It became a brilliant demonstration of a deductive science in which theory predicts and experiment confirms. This is the story, almost fantastic, that describes the discovery of the earliest erectine, Java Man, in which Eugène Dubois took Ernst Haeckel's prediction of what the missing link between apes and humans ought to look like and where it might be found and then dug the evidence up. The episode is as close as paleoanthropology will likely ever come to an *experimentum crucis*. It seemed to confirm the evolutionary theory of human origins as fully as Eddington's 1919 measurement of bent starlight did Einstein's theory of relativity.

The discoveries of *Homo neanderthalensis* and *Homo erectus* between them contain the basic templates, and tropes, of discovery in the field. They also illuminate the ways in which field and theory might not only direct but distort one another. While the bipolar nature of scientific inquiry helped to sort the genuine from the phony, it also meant that some legitimate finds might be discredited outright or disputed for decades, as happened with Raymond Dart's announcement of the Taung child, the first australopithecine, and it meant that some frauds might be accepted, even lionized, as happened with Piltdown Man, a putative missing link between the erectines and the Edwardians.

Links Lost

That gaps persisted was equally frustrating and useful. They meant that knowledge was incomplete and hence theory unstable. But they also identified targets for research—they pointed the field to where it needed to go and what it needed to find. Paradoxically, the still to-do helped make the already-done believable. It made something like normal science possible.

Still, the existence of the gaps needed to be explained as well as filled. When the Chain of Being prevailed, they best were understood as Leibniz argued, as lapses in the existing state of knowledge. The species existed but were not yet found; the gaps were failures in human understanding not breaks in the plenum of creation. The evidence was there. Eventually human ingenuity would discover it. Everything had its place, and there was a place for everything.

As geology matured, the recognition grew that some gaps were permanent. The material evidence had eroded away. In a common image, it was held that while everything had been written in the book of nature, over rough eons some pages had been torn out and would never be restored. Since nature was continuous, those gaps were only apparent, an artifact of spotty preservation. The breakdown lay in the ability—or in this instance, the inevitable inability—of science to recover that lost lore. This in fact was Darwin's position, and it became the received wisdom. Nature was continuous and graduated, just as the Chain insisted, but like a library some books had disappeared from the shelves.

Even so, for the science to fulfill its mission those gaps had to be bridged, and for the grand narrative to be complete, the story had to be told despite missing episodes. The temptation was enormous to fill what the mind said—what the prevailing wisdom of the day insisted— had to be there. If missing fossil links weren't found, they might be hypothesized, which could slide into invention. If history failed be- cause it could not cross geographic gaps, then something had to span those seas. If narrative sputtered because character and events were lacking, then some medium would fill the vacuum. For the gap in data, there was theory. For the gap in space, there were hypothetical lands. For the gap in story, there were traditional formulas, the miss- ing links and land bridges of narrative.

Where did humans reside in the Chain qua tree? Pope placed them in the "isthmus of a middle state"—a lovely metaphor but not much of a guide to research. Instead, as happened with history, moral philosophy as the intellectual predecessor to anthropology took its data from the Bible and the classics of antiquity, and then, thanks to the Renaissance's voyages of discovery, it added travelers' tales full of both fact and fancy to the stew. The number of peoples to organize grew uncomfortably.

Then Linnaeus attempted to codify the results. In the first edition of *Systema Naturae* (1735) the definition of the human was unstable, and Linnaeus threw in references to satyrs (from classical literature) and "tailed men" (from travelers' tales) as possible collateral members. By the second edition (1740) he had recognized subspecies according to the major historically emphasized racial divisions (European, American, Asian, and African), along with the feral and the mon- strous. The critical change came with the 1748 sixth edition, with the injunction *sed homo noscit se ipsum* ("but man learns to know him- self"), the placement of humans among the anthropmorpha, and the identification of reason—of which this was the self-identified age— as the distinguishing characteristic. He confessed that on morpho- logical grounds, based on "the principles of Natural History," he could find absolutely no other generic differences. Already the sense triumphed that the distinguishing trait between modern and archaic humanity would lie in behavior, not in morphology. The soul had secularized into reason.[20]

In the 1758 edition, which set the standard for systematics, he placed humanity among the primates and distinguished two species: *Homo sapiens* and *Homo troglodytes*. The latter lived in caves, occupied Ethiopia and Java, and were intermediate between apes and people. They are "most nearly related to us," Linnaeus concluded, and "affirm to be former rulers of the world, but they were deposed by men"—an eerie premonition of prior ancestors, their fate, and the probable location of their remains. Those understandings, however, were the ancestral stock from which have spawned all successive theories.[21]

Evolutionary thinkers unhesitatingly applied their doctrines to *Homo*, and Linnaeus's *scala* morphed into Haeckel's ever brachiating tree trunk. Haeckel himself regarded the great anatomical difference between humans and apes as bipedalism, and the way extremities were "differentiated accordingly," but behavior was the most telling distinction in the form of "articulate speech" and "higher reasoning power." When he forecast that the twenty-first stage in the Chain of the Animal Ancestors of Man would be found in south Asia, and then Eugène Dubois found the bones as predicted, the search for other missing links began in earnest.[22]

No one could claim that Dubois's quest had been easy, yet it had, by most standards, been quick. Other fossils came slowly; most belonged to the growing clan of European Neanderthals, as searchers entered cave after cave. But the discovered phylogenetic links were few. Not until 1925 did Dart describe the Taung child, and not for another year did the fabulous Pleistocene treasures at Zhoukoudian become known. Another decade would pass before these were confirmed and generally known, and still more decades until the discovery of early hominins became a serial enterprise. That left a large gap—not only in the phylogeny of hominins but in the scholarship of hominin study. Human nature moved to fill the void with false finds, by seeing more than was there or inventing what refused to show itself.

Links Invented: Fossil Figments of the Paleoanthropological Imagination

With the expectation that paleontological gaps were links that were missing but ready to be found, for some practitioners the urge became overwhelming to fill them in. Some created links were simple errors, some outright forgeries. But all were accepted because they were believable, and they were believable because they fit the prescribed traits of what the intermediary forms were supposed to look like, and they satisfied intellectual (and cultural and personal) desires. Perhaps none, however, had so absurd a life cycle as *Hesperopithecus harold-cookii*, popularly renowned (or infamous) as "Nebraska Man."

The fossil had everything a missing link should have. It displayed just enough dentition to allow for morphological ambiguity in what was the hard data. It appealed to a field collector, an avid amateur with enough connections to bring the artifact to suitable attention. It attracted the interest, and authority, of a major researcher with powerful institutional ties that could add further scholarly heft and publicity. Its setting was perfect: the same Nebraska badlands that had yielded *Eohippus*, the missing link for the equids, in the home state of a prominent populist politician and critic of evolution, William Jennings Bryan. And its timing was brilliant. It emanated from a boisterous America that lacked any evidence of early hominins, had in fact a history of official hostility toward claims for such creatures, and was as hungry for international prominence in fossils as in battleships or stock markets—and was preparing for a media circus on evolution, the Scopes trial. That a man of the people, Harold Cook, rancher and part-time paleontologist, should discover in the American backcountry a missing link that could command world acclaim was as though George F. Babbitt had found a buried Aztec treasure in Zenith. No small-town booster could have invented a sweeter deal. And with the debate over evolution commanding national attention, no fossil ever found a wider and more politicized audience. It all seemed too perfect.[23]

And so it was—the tooth was worn and far from definitive, but caution went to the winds when Bryan challenged Henry Fairfield Osborn, president of the American Museum of Natural History and a prominent publicist for evolution, to find the missing link between apes and people. Osborn urged Bryan to study "the simple archives of Nature"; days later, the tooth arrived at the museum. "Speak to the earth and it shall teach thee," Osborn had intoned from the Book of Job. The Nebraska tooth was the reply. Osborn named it after its discoverer, *Hesperopithecus haroldcookii*.

Osborn made casts and distributed them to colleagues, including prominent Europeans. Inevitably, there were both skeptics and zealots. The project took a theatrical turn when the *Illustrated London News* ran an artist's reconstruction of a Hesperopithecine family at work on the Pleistocene Plains. Amedee Forestier modeled his image on Dubois's interpretation and the idealization of Dubois's Java Man, making Nebraska Man into the earliest of America's immigrants and the first of its prairie pioneers. Soon, however, the silly became the mocking.

While both sides focused on missing links, one accused the other of dismissing what nature revealed and the other charged that its rivals were revealing what nature had never produced. Shortly before the 1925 trial, Osborn had asked Bryan, "What shall we do with the Nebraska tooth?" He answered his own question by affirming, "Certainly we shall not banish this bit of Truth because it does not fit in with our preconceived notions." He was thinking of fundamentalists who challenged any evidence in favor of evolution. But he might have reversed his audience and pleaded not to accept as truth whatever fit into preformed notions, whether they were drawn from Genesis or Haeckel. Early on Osborn had proposed that the link ought to be named *Bryopithecus*, "after the most distinguished Primate which the State of Nebraska has thus far produced." Afterward, the antievolutionist John Roach Straton returned the compliment by proposing that the toothsome creature be named *Hesperopigdonefoolem obsorniicuckoo*, after the country's second most prominent primate.[24]

More methodical studies suggested that perhaps the tooth belonged to an anthropoid ape rather than an ancestor of *Homo;* and then fieldwork on the site in 1925 pointed to the possibility that it might come from a Pliocene peccary; and on the very eve of the Scopes trial, when the tooth might have bitten into antievolutionists, it quietly disappeared from the political scene. Late in 1927 W. K. Gregory, to whom Osborn had handed over the technical study of the tooth, confirmed its correct status as a pig's tooth with a retraction in *Science*.[25]

If Nebraska Man tacked close to populist parody, threatening to become the Billy Sunday of American paleontology, Piltdown Man veered into intellectual pathology. The one was foolish, the other, malign. The Piltdown hoax is surely the most prominent of paleoanthropological frauds, as well as its longest-running intellectual soap opera.[26]

In early 1912 an amateur collector of antiquities, Charles Dawson, presented to Arthur Smith Woodward, a prominent paleontologist with the British Museum, some skull fragments that he claimed to have gotten four years earlier from a workman at a gravel pit near Piltdown, England. He and Woodward then visited the site and recovered some additional artifacts—a cranial bone, half a mandible, and two teeth—which Woodward then assembled into a plausible skull. In December 1912 Woodward announced the discovery of Piltdown Man to the Geological Society of London.

Woodward believed that the composite was a missing anthropoid link. The jaws resembled those of an ape, while the skull and teeth were those of a modern human. The partial skull offered a perfect complement to Dubois's discovery. Java Man had a small brain, Piltdown Man, a large one. Eurasia thus established the hearths for the evolution of early hominins, as prevailing theory predicted. That a big brain had come early supported British predilections; and that it was found in East Sussex stroked British vanity. Even the scenario of discovery, almost an echo of Neanderthal, mimicked expectations. From the outset Piltdown Man thus promised to be canonical. Its morphology placed it between apes and humans, its location sited it close to research centers, and its manner of reconstruction conformed to normative patterns. In brief, it fit so many gaps in knowledge, ambition, and desire that it might have been explicitly contrived to do so.

And it was. What is most remarkable is how long it took to identify the artifacts as fraudulent and the reconstruction as fanciful. Not until 1953 did a major review expose the episode as a hoax. The reasons for its persistence were several. It begins with the character of the fragments, which were broken in ways that left their reassembly ambiguous. As more than one observer noted, deliberate malice could not have shaped them in a more cunningly exact way to feed into presuppositions. By enlisting the passionate support of Woodward, and through him much of the British paleontological establishment, the Piltdown skull seemed to speak with the chorus of authority. Others were sucked in as well, including such luminaries as Pierre Teilhard de Chardin, Arthur Keith, and even Arthur Conan Doyle—who, himself, seemed in transit between Sherlock Holmes and the Lost World. There was, too, an undeniable element of nationalism; it is no accident that most early skeptics worked outside Britain. And the gap in the chain was large: that *Homo erectus* was discovered in the early twentieth century only accented the need for further links between it and modern humanity. Announced in 1912, Piltdown Man promised to plug that hole.

In truth, the skull was discovered to be a clever artifice that combined an orangutan's jaw, a medieval man's parietal bone, and two chimpanzee teeth. The jaw was broken at exactly the place where it would join the cranium, and thus allowed for a wide range of potential refittings. The teeth had been filed to make them resemble human dental wear. The partial parietal hinted at a large cranium without specifying its dimensions. The bones had been chemically stained with iron and chromic acid to give them a proper patina of age.

But if the reassembly of those pieces seems improbable, no less so is the assembly of characters involved. The perpetrator has never been confirmed, although the primary pigeon was surely A. S. Woodward. The whole field ended as stained as the doctored bones, for the episode deflected the research interests of a major scientific power, hardened sentiments against discoveries in Africa (such as the Taung Child), and distorted public expectations (Clarence Darrow even introduced Piltdown Man as evidence into the Scopes trial, perhaps the ironic equivalent of a double entendre). The real testimony, however, is to the intellectual power of the missing link as an organizing concept and a prod to field research.

Links Invented: Geography Recapitulates Phylogeny

The distribution of species, whether living or fossil, was a matter of geography as well as geology. In order for species to evolve and radiate, evolution required ample servings of both time and space. Thus, the flip side to evolution was biogeography, and it is no accident that the greatest practitioners of that field, such as Alfred Russel Wallace, should also have been among the founders of evolutionary theory. Some species are joined because of spatial arrangements and some because they share a common past. Interestingly, this meant biogeography had spatial gaps as much as phylogeny did temporal ones. Its missing links were lost lands.

These gaps were filled by land bridges, which were the product of scientific guesswork, and occasionally by wish-fulfilling error that in some circles hedged close to hoax or self-delusion. Land bridges built on a long Western tradition of imagined islands and sunken continents in the unexplored Ocean Sea. As biogeography's rosters swelled, connecting species that today are separated not only by formidable straits but by gaping oceans, the number of putative terrestrial spans proliferated and their sagas became more improbable, as they were assumed to rise and fall like the tides of Fundy. The Arctic Atlantic had the land bridge Archiboreis; the mid-Atlantic had Archatlantic to connect the West Indies and North Africa; the south Atlantic boasted Archhelenis to join Brazil and southern Africa. There was an Archinotis to bond South America to Antarctica. If, for poets, the past had been a time when heroes and giants walked on Earth, biogeographers seemed prepared to imagine the lands they had walked on. Everywhere, it seemed, land bridges had connected to everywhere else.

The most spectacular was Lemuria, the brainchild of Philip Sclater, who puzzled over the mammals of Madagascar, specifically lemurs, which were found in Madagascar, the Mascarene Islands, and India, but not in Africa or Asia Minor. Sclater postulated that some landed arch must have spanned the Indian Ocean, a missing link of biogeography that he termed Lemuria. Haeckel boldly elaborated the idea into a continent-scale landmass that had once stretched from East Africa to East Asia, from "Madagascar, Abyssinia" to "Sunda Islands, Further India." It was in this putative paradise that early humans had evolved. Subsequently Lemuria had "sunk below the waves of the Indian Ocean" and taken the missing links of evolution with it. Subsiding lands thus joined eroded strata among the missing pages of the book of nature.[27]

All in all, it was a fantastic conception, but one true to the fossils and to the evidence of species distributions. In 1944 G. G. Simpson, the doyen of paleontology, was still invoking land bridges as part of the modern synthesis of biology. It was "[a]mong the plainest infer-ences" that mammals had formerly crossed to regions "that are now impassable to them" and that "expansion and contraction" were reali-ties of the geologic record. Simpson then brought the grand themes of the Pleistocene together by proposing as a model for such action the "excellent descriptive analogy" afforded by ice sheets; as an exem-plar of how fauna might behave, the "enlightening" exemplar of the mastodonts; and as mechanisms, several theoretical models of land bridges transporting, filtering, and barring. There the subject remained until the theory of plate tectonics pulled the concept of far-spanning land bridges permanently into a subduction zone.[28]

Links Written: Missing Tropes and Lost Texts

That left the question of gaps in the telling. The evolving concept of the Pleistocene might frame the story of human origins, but it could not organize its narrative, which had to look elsewhere for templates. Even Dubois's Java Man or Britain's Piltdown Man had to fit not only into categorical chains and typological trees but into an interpretive chronicle, and narrative had its own necessities, among them a continuity of episodes leading to a conclusion. Those missing links of literature also had to be filled.

One solution was to look to tradition. In the same way that *Homo troglodytes* updated classical allusions into Enlightenment codes, and that Lemuria and Archatlantic were successors to Terra Australis (a great southern continent that logic demanded to balance out the landmasses of the northern hemisphere), so the events of hominin evolution might seek to dress up in more rational garb and with more plausible incidents the formulas of classical storytelling. And just as evolution echoed other developments in the larger culture, so it would prove with narrative. It helped that during the decades after Darwin, when evolution became the dominant scientific theory to explain human origins, literature and history had turned to narrative and did to it what evolution had done to the Chain of Being.

The interplay was mutual, since the new creation stories needed a more robust plot, while the discoveries of the sciences poured new grist into their literary mills. The classics of the period—from that rush of early books by Huxley, Lyell, Darwin, and Haeckel through the twentieth century's successor tales by Arthur Keith, Elliott Smith, Frederic Wood Jones, Henry Fairfield Osborn, and William King Gregory—corresponded precisely with both the narrative style of literature and the search for missing links. Science suggested new methods of textual analysis and narrative supplied what field excavation could not.

In the late 1970s Misia Landau began her observations, mentioned previously, that the classics of paleoanthroplogy from this era emulated the structure of Russian folk stories as described in Vladimir Propp's *Morphology of the Folktale*. Revealingly, Propp's treatise was published in 1928, making it contemporary to the narratives Landau analyzed. Their commonality took two forms. One, Propp treated stories as biologists did species, or as he put it, "according to their component parts and the relationship of these components to each other and to the whole," another extension of the comparative methods that had proved so powerful for anatomy and embryology. The second is the way their subjects used narrative as a medium of expression. It is not that evolutionary thought flowed from literature, or that literature descended from biological science, but that both shared the assumptions and aspirations of a common intellectual syndrome. The nineteenth century was, after all, the grand era of evolutionary theory, narrative history, and the novel. Tolstoy's *War and Peace* is as much a model for describing the world as folktales from the taiga. In explaining human origins the principal paleoanthropologists of the time spoke with the venacular of their time.[29]

What such texts need is character, conflict, and climax. They need a protagonist who is capable of change and who possesses agency. A character who remains unmoved may be a superhero, but he is not a figure of more than comic book or entertainment interest. Likewise, a character who can only accept passively the blows of fate engenders scant interest and less sympathy. When confronted with a challenge—a villain, a dark double, a hostile setting, a threatening predator or force of nature—the protagonist must have the capacity to respond. He must be able to choose, and the cost of that choice must be real. The conflict may be singular or successive (in the late nineteenth century, subplots rich with characters and contests grew like a spreading chestnut tree). But they must end in a climax, and this resolution will shape everything that has happened before, because they will all lead to that end. To paraphrase Anton Chekhov, If in the first act, you introduce bipedalism, in the next act you must use it.[30]

In her exegesis Landau compares the anatomy of the hero folktale with the classics of paleoanthropology. There is a mysterious birth, gifts from a "donor," a succession of tests, a triumph, and a return, which may end ironically (or tragically) with hubris, nemesis, and personal destruction. The particulars of the evolutionary tale change, as they do in traditional tales. The challenge may be glacial climate or short-faced bears; the series of challenges and choices may be shuffled from bipedalism to encephalization, according to taste and theory; the prime mover (or "donor" in Propp's and Landau's eccentric phrasing)—the motive force that carries the action—may be internal or external, but they all proceed stage by stage, met contest by met contest, to a climax, which is modern humanity. In its very structure, by its sequential unfolding as much as by its theme, the narrative explains who we are.[31]

As with historical narratives, however, the movement may go down as well as up. Most literature from the late nineteenth century and the early twentieth was progressive; it depicted a world that despite its trials went onward and upward. But the very determinism of that aesthetic and thematic design might trend downward as well, and the literature of the era was full of implacable laws of nature working toward dystopias. After two world wars and a global economic depression, the old narrative did, it seem, get inverted, with humanity's origins bestowed by a dark donor, or at least was lost in mist. That transition describes nicely the arc from Darwin to Dart. For all his skepticism of Providence, Darwin saw an evolutionary progression toward complexity that led to a hopefully ambivalent humanity. Raymond Dart and his successors saw the transition to humanity as leading to a savage predator and cannibal, "a loathsome cruelty of mankind to man," so intrinsic to humanity's evolutionary success that it constitutes "one of his inescapable, characteristic and differentiative features." The narrative structure could point either way.[32]

Then in the postwar period a modern synthesis slowly gelled that altered both the science and the literature and how they might interact. More than a question of what kind of data science had, if it wished to tell creation stories the style of narrative would have equal say in shaping the outcome. The discipline might disdain discussion of missing links, but story as a form of description and synthesis had a power beyond its capacity to fill in evolutionary blanks. It could

endow meaning and moral drama. Until the mid-twentieth century
the story told took itself the form of a story.

Method, the Modern Synthesis

The twentieth century opened with the first tremors of a wholesale
rehabilitation of intellectual culture—the largest since the Enlightenment. Ernst Haeckel, always at the center of the action, deftly illus-
trates the transformation. Between 1901 and 1914 he revised and
republished virtually all of his major works, from *The Riddle of the
Universe* to *The Evolution of Man* to *Art Forms in Nature* to *The History of
Creation*, the last issued on the eve of World War I. But already such
summas of metaphysical history seemed forced and archaic. In the
world of ideas they resembled the colonies artlessly assembled from
the partition of Africa or the last gasp of geographic exploration preparing to die on the ice of Antarctica.

The tree went the way of the Chain, and the concept of missing
links wandered into the disreputable realms of popular culture and
Sunday supplement journalism. The Chain had reflected its world,
and when that felt apperception—the ways in which nature and culture seemed to reflect each other—no longer made sense, it found a
new avatar by adopting history as an informing principle and as an
explanatory medium better suited to a world that seemed most aptly
characterized by growth and expansion. By the early twentieth century that reconstructed world was also breaking apart under the
blows of revolution, war, economic crisis, and a metaphysical reconstitution of humanity's place in the great scheme of things and a
reconsideration of how we might express that understanding. However imposing its arboreal architecture, the intellectual heartwood of
the Tree of Life was rotting away.

In 1905, as Haeckel published his *Last Words on Evolution*, Albert
Einstein defined the *annus mirabilis* of modernism with breakthrough
works on randomness (Brownian movement), the special theory of
relativity, and the photoelectric effect. Two years later Pablo Picasso
painted *Les Demoiselles d'Avignon* (1907), announcing the most radical

alteration of perspective since the Renaissance. A cascade of novelties, many radical in their output and assumptions, rippled over the scene in field after field, as the old order died on its feet and a new one slowly arose. Bit by bit Haeckel's comprehensive *Weltanschauung* fell to pieces, with his romanticism tainted by absorption into Nazi propaganda, his biology broken by the new genetics, and his narrative fractured by the demise of historicism. What made him a genius in his own time condemned him to future scorn.

The modernist temperament came late to those disciplines most relevant to the study of the Pleistocene. It came to anthropology in the 1920s and 1930s essentially through the work of Franz Boas and his students, and became institutionally dominant in the postwar period. It came to the life sciences in the 1940s and spread pervasively only after World War II. It crystallized for the earth sciences in the late 1960s with the theory of plate tectonics. Each challenged the claim that history could best synthesize knowledge and make predictions about the future. Cultural anthropology broke up into the study of discrete societies, each with its own relativized moral order, a collectivity not joined under a grand narrative of evolution. Life science reasserted, with a more sophisticated genetics, the original formulation of Darwinian evolution as open-ended rather than an implacable progression from incoherent homogeneity to increasingly coherent heterogeneity. Earth science shed its complementary (if cartoonish) vision of a planet that like a rocky embryo evolved from a simple molten blob into a complex integration of seas, lands, minerals, and mountains. Modernists felled the World Tree as they hoped to raze overgrown cities.

Unlike previous eras, this one could point to no single expression—no iconographic diagram, no master metaphor—to summarize its meaning. Instead, the old forms were ruthlessly dismantled or fell to the wayside, their ruling narrative broken up not unlike the process of decolonization that occurred at the same time. This left adrift the question of how to express the new understanding; and this time art, literature, and philosophy offered no ready replacement. All turned inward, more intent on contemplating themselves than an outside world.

The arts challenged not only the Old Masters but the prospects of surrogates. Cubism, Dadaism, fauvism, constructivism—none readily lent themselves to a representation of modernist nature in the mode of a chain or tree, and, in fact, they rejected the very idea of art as representation. *Guernica* or *The Bride Stripped Bare by Her Bachelors, Even* might be canonical works for the art of their day, but they are not the stuff by which one might visually join biology and philosophy the way its frontispiece did Erasmus Darwin's *Temple of Nature* or his poem of evolution, "Zoonomia," or Alexander von Humboldt's might depict the life zones of Mount Chimborazo.

Philosophy came both earlier and later into the mix. American Pragmatism had, almost uniquely, seized on the element of chance inherent in Darwin's original edition of the *Origin* to argue for an open, pluralistic universe, dismissing Hegelian and Haeckelian "block-universes," as William James described them. That tradition got swept out to sea by the influx of intellectual immigrants beginning in the 1930s. In its place appeared, for the more metaphysically disposed, existentialism, and, for the more scientifically inclined, positivism. What all shared was a suspicion regarding assertions of a moral order in nature toward which some historical process might trend. That fascism and communism both held ideologies informed by such sentiments made one an intellectual enemy during World War II and the other a rival during the Cold War. To argue that such ideologies only represented the political expression of nature's order seemed mad, or worse.

The critique found its modernist voice with Karl Popper's *The Poverty of Historicism*. Begun in the mid-1930s and revised over the next twenty years, the treatise sought to show that "historicism is a poor method—a method which does not bear any fruit." By historicism Popper meant an approach to the social sciences "which assumes that historical prediction is their principal aim, and which assumes that this aim is attainable by discovering the 'rhythms' or the 'patterns,' the 'laws' or the 'trends' that underlie the evolution of history." In effect, the grand scaffolding of ideas that had looked to history as a law of nature, and for which evolution had been the triumphal exemplar, was a scientific sham—and a morally repugnant politics. (Popper dedicated his treatise to the memory "of the countless men, women and children of all creeds or nations or races who fell victims to the fascist and communist belief in Inexorable Laws of Historical Destiny.")[33]

Historicism's intellectual disgrace mattered because, by now, only those disciplines that could claim standing as "science" had legitimacy, and science was demonstrated by its ability to make testable predictions. The collapse of historicism barred any subject informed by history from being a science. History could be explained after the fact, but not before. The historical record was less an immense tree that grew predictably from seed to spreading maturity than it was a game of chance full of quirky regularities, none of which was more predictable than a reckoning of odds. It was impossible, for instance, to specify what properties a missing link might have, or to identify in advance where in a phylogeny it might be found, or what the next, future link might look like. History, and its literary expression, narrative, were not scientific, and so did not deserve the attention that true sciences did. Only science could yield positive knowledge—hence the passion for positivism, and the origin of the term. Narrative thus joined missing link as phrases that vaguely reeked of the unprovable, the naive, and perhaps the vulgar.

If the tree was the visual expression of a providential history, narrative was its literary manifestation. Romantic literature was, as an art form, the perfect complement to narrative history: they both shared the same rhetorical forms, similar conceits, and a reinforcing recourse to unfolding plots for which the end determined the means. They were both teleological. The complex phylogenies displayed by the fossil record showed a comparable design in the progression of their plot as the novels of Charles Dickens or Alexandre Dumas. The novel was a historical literature as the grand history was a literary genre. They intertwined, growing from common roots of art, philosophy, and history like Yggdrasil under the hands of the three Norns.

Unsurprisingly, the disintegration of historicism coincided with a turning away from, and even outright repudiation of, narrative. Because it required aesthetic closure, an end to which the text pointed, classic narrative carried within it the seeds of teleology—that's what made it work as literature—and what made it suspect for writing anything that purported to be scientific and traced changes through time. The modernist revolution in literature broke down the standing of narrative just as philosophy did historicism. It challenged the authority of an omniscient narrator, which is to say, a universal perspective; it questioned the sequential unfolding of events, accepting time as relative and uneven; and it denied the purposeful movement of story from origin to end like the stages of an embryo. The modern synthesis of history lost its claim to universality, became lumpy and jumpy, and allowed analysis to replace synthesis.

Literature accepted missing links in its plots; it might even celebrate those gaps. In the narratives of the nineteenth century, story had built up, event by event, like dressed stones in a load-bearing wall. The higher the wall reached, the wider its base had to spread. In the narratives of the twentieth century, story leaped from node to node, its weight borne by a grid of beams. Like a skyscraper, it could soar from a narrow base, and each wall did not have to carry the weight of all those to come. A William Faulkner novel does to narrative what a Georges Braque Cubist painting does to mathematical perspective. Joan Didion stands to Charles Dickens as a steel-girder grid to a load-bearing wall of bricks.

Those differences also describe deftly the distinctions between the histories of their eras and the phylogenies constructed for the evolution of species, especially of hominins. In the Chain and the tree species were connected by link or trunk, and if any were missing, it was possible to predict what they would look like and where they might be found. In the late twentieth century every hypothetical phylogeny differed, connections were dashed lines, and gaps were question marks. While truer to positivistic science, such reconstructions did not address—could not address—the deeper yearnings of the sustaining culture. Instead, those longings still turned to narrative, and narrative slid into popular culture, or imploded into national or ethnic creation myths, or simply dissolved into personal memoir. Narrative as a means of analysis and synthesis remained tainted for high culture and scorned by fields that aspired to scientific legitimacy.

This solved, nominally, the question of doing science. Acquiring data was split off from the task of organizing or interpreting those findings; the number of points suitable for a scatter diagram proliferated. The solution left unsettled, however, the matter of how to connect them, or whether a regression line would prove as satisfying as a narrative arc. It was possible to segregate fact from story, but if the larger purpose was not the amassing of fact but of meaning, and meaning came at least partly from the moral drama inherent in story, then what, literally, was the point of all those points? After all, the ultimate purpose of inquiries into humanity's place in the scheme of things had always been moral: to identify what makes us human; to say who we are and how we should behave.

It had been the special task of Pleistocene studies, and especially of those who worked on the history of ancestral hominins, to weld those implicitly metaphysical questions to scientific data and to suitable means by which to express those findings. When historicism failed, they turned to positivism, which broke the links that had fettered research to Chain and Tree. Instead of purpose, one had process (and the New Archaeology based on that premise even called itself "processual archaeology"). The medium—the method—became the message. One had the simple doing. This began as pragmatism but shunned the pragmatist's ardor for vision and openness and segued into a tight-fisted positivism that scrapped all but the narrowest confirmations offered by science.

Like a Hemingway code hero, one had craft, an alloy of grace and skill displayed in the face of (or quiet defiance of) a larger world that one could neither know nor express. One could hunt for elusive fossils as Hemingway and Pop did the African kudu in which the search—the technical methods that led to more explicit accretions of fact—was the task, which might hopefully end with the trophy head of a hypothesis that could organize those bones into significance. With ruthless dismissal they shunned the fancy words in the form of the elaborate narratives that had been the triumph of their predecessors. Instead they wanted plain facts and valued the reductionist skills that produced them.

Even those who still recognized the need for narrative found the Haecklean tree anathema because it led, branch by branch, to a hominin apex. Rather, that towering tree more resembled a rough, harshly pruned shrub, less a redwood than a blueberry. "Evolutionary 'sequences' are not rungs on a ladder," insisted Stephen J. Gould, "but our retrospective reconstruction of a circuitous path running like a labyrinth, branch to branch, from the base of the bush to a lineage now surviving at its top." The narrative lay not in the saga of the sapients but in the minds of those sapients doing the research. Here lay both the power of science shorn of cultural associations and the poverty of the positivism that filled the resulting vacuum. The science was better—no one doubted that. It could excavate more accurately, age-date recovered specimens more precisely, extract more information from artifacts, and propose more plausible reconstructions from fragments. It could shear away the more outrageous speculations. In all this it seemed to demonstrate the power of the positivist philosophy that prevailed. Method became itself the meaning.[34]

In fact, the new age did not adhere to a genuine philosophy of positivism but to a bowdlerized version that had degenerated into a latter-day Baconian fever in which the collection of data, which the science could do, substituted for answering the motivating questions, which it could not.

Extinct Links and Eternal Verities

Across Eurasia Pleistocene hunters had assembled the bones of mammoths, a creature they knew, into hide-covered shelters. Then both prey and predator disappeared, and when the massive bones were from time to time encountered in subsequent millennia, they were waved aside as the relics of mythical creatures or ancient giants. In the eighteenth century, however, another category of hunters found mammoth bones, and they tried to fit them into another kind of *domus*, this time an intellectual one. By now, however, they no longer knew of any such animal. They applied the local Russian term, *mamont*, to describe the putative creature that bequeathed the relic bones.

In 1728 Hans Sloane penned the first scientific study, based on fossil teeth and tusks. He assumed they must have come from creatures destroyed by the great deluge of Noah's time. Steadily, quickening with the second age of discovery, more bones came on the scholarly market. J. P. Breyne reported on additions from Siberia. Mark Catesby recorded teeth collected in North America. But no living embodiments were found, despite the enlarging domains of exploration, and in 1796 Baron Georges Cuvier, a founder of comparative anatomy, proclaimed the obvious. The mammoths were not elephants (somehow transported to permafrost); rather, they represented a species that no longer existed. Three years later Johann Blumenbach assigned a binomial name, *Elephas primigenius*, thus emplacing them within the Linnaean scala naturae, and Thomas Jefferson, pondering the meaning of pieces excavated from Big Bone Lick, Kentucky, tried to insert them into the Great Chain. "The bones exist: therefore the animal has existed," and since "[t]he movements of nature are in a never ending circle [*sic*]," a species that "has once been put into a train of motion, is still probably moving in the train."[35]

But it wasn't. Buffon resolved the dilemma by imagining a series of catastrophic extinctions and fresh creations, each grander than the one before. Jefferson's vision was more inclined toward that of Pope. "If one link in nature's chain might be lost, another and another might be lost, till this whole system of things should evanish by piece-meal." That prospect was too weird to contemplate, so he assumed that "if this animal has once existed, it is probable . . . that he still exists." Such was the logic and economy of nature. It was not, however, the logic of history. The animals had in fact disappeared, one by one. The woolly mammoth was the founding, astonishing emblem of that wholesale vanishing. It soon became the canonical icon of the Ice Age, as people have once more picked up mammoth bones, and have once again arranged them into an edifice that people will occupy. This time the ribs and femurs and tusks assemble into a *domus* of understanding. The mammoth was the first link of the Chain to be broken, and among the first of a proposed new order to replace it.[36]

In its fullest form the Great Chain of Being expressed a cluster of principles that together gave the how, the what, and the why to the scheme of things. The metaphors of the chain and the tree fused them into a powerful allegory cum theory, but the parts could change separately, and did. The most mutable was the *how*, expressed in the principles of gradation and hierarchy. This was the domain of science, and the principles underwent a relentless hammering as empirical discovery and experimentation recast a nature that was understood to move slowly and incrementally into one that was lumpy and jumpy. So overwhelming and spasmodic was the outcome that the Chain could no longer hold all the pieces into a coherent ensemble, whether by links of taxonomy or of history. What had been immutable became not only mobile but unmoored from its metaphysical end points. The links became data points on a scatter diagram. They persist only as metaphoric fragments in such expressions as food chains or trophic chains.

The *what* was the principle of plentitude, which stated that the universe was as full as it could be, and its very plenum was the best possible arrangement the world could know. This sentiment has survived intact and has even acquired a scientific discipline, conservation biology. But the belief has extended tendrils into most of life science and environmentalist enthusiasms, as it insists that every link is equally valuable and the loss of any single part can undo the whole.

The *why* was the principle of sufficient reason, what Aristotle had called a final cause. It held that the universe was its own justification; baptized into religious lore, the material world represented the Creator in his fullness and splendor. In recent decades the doctrine has endured as variants of environmental ethics and, whether acknowledged or not, in nominally scientific arguments for protecting species and ecosystems. It is present in philosophies of deep ecology and in doctrines of nature's intrinsic value—and in condemnations of human hubris before the inscrutable presence of life.

The metaphors have wasted away, so that today they stand to complex systems of ecological science like split-twig figurines left in caves. They can no more explain the principles of physical nature than an icon of St. Elias can halt a forest fire. But their ultimate principles have always spoken less to the how than to the what and why, to a sense of transcendence in nature or a permanence of principles in the constitution of the world. In this persistence they have proved as immutable as the Great Chain insisted they were.

In the early twenty-first century it was no longer unthinkable, as it had been in the early eighteenth, to imagine missing links—not links yet to be found but links once known and now broken permanently from the Chain. Today they seemingly tumble from biotas like snow from the sky. They are taken as signs of disequilibrium and malfeasance: the scheme of nature violated and broken. Their loss threatens practical human existence, since those species do ecological work and harbor natural wealth; but more, their destruction testifies to a fundamental disorder that threatens to rot away the core of humanity's moral agency. It is not merely unwise but unethical to allow species to die out, much less to harass them into extinction. In a celebrated passage the wise man of modern American environmentalism, Aldo Leopold, observed that the first rule of intelligent tinkering was to save all the pieces. Pleistocene humanity, it seems, tossed the loosened screws and springs aside.

It remains for contemporary humanity to put the residual pieces back together and, where necessary, to rebuild them from scratch. In recent years proposals have been floated to return a patch of North America's Great Plains to Pleistocene-equivalent species and to establish a suite of "Quaternary Parks" stocked with the similacra of vanished species, a means of "restarting evolution of at least some of the lost lineages." These would include elephants as approximations of the lost mammoths—what paleontologist Paul Martin, best known for advancing the "overkill" hypothesis to account for the Pleistocene-ending extinctions, has called "resurrection ecology."[37]

The most spectacular expression is a Siberian scheme dubbed Pleistocene Park. Conceived by Sergey Zimov, sited in northern Yakutia, the project aspires to re-create the landscape that existed in the late Pleistocene. The premise behind the restoration is that since people killed off the creatures, helping set in motion the upheaval, restoring the lost bison, reindeer, horses, elk, musk oxen, antelopes, and assorted carnivores like the Amur leopard and Asian lion, would allow the land to heal, since those lost megafauna had shaped the land as much as climate had. The supreme restoration, however, would be—what else?—the woolly mammoth. The scheme would require extracting DNA from the carcasses of frozen mammoths, an experiment that may never happen. The exercise in rewilding or, more accurately, in reincarnating deftly merges both the geographic and the genomic narratives, and it would likely mean restoring a megadose of unintended irony.[38]

The reserve is a scant eight square miles, although it is surrounded by a research preserve and a large buffer zone. Yet if the land base is small, the impulse is not. It harks back to those inchoate instincts and ideas that Plato gave form to in the *Timaeus* but which surely date from the earliest imaginings of Pleistocene humanity. So while the project advertises itself as a scientific experiment, and the hows matter because technical expertise will be essential, the real motives are the what and the why. It seeks to restore the principles of plentitude and sufficient reason, to reassert the moral order of creation, a conception that the West first crystallized into a Chain of Being. Amid such ideas we are not dealing with eternal links but with eternal verities and utopian yearnings, here reincarnated into a prelapsarian Pleistocene.

Chapter 7

New Truths, Heresies, Superstitions

History warns us . . . that it is the customary fate of new truths
to begin as heresies and to end as superstitions.

—T. H. Huxley, *The Coming of Age of the Origins of Species* (1880)

THE NEANDERTHALS have challenged commentators from the time of
their initial discovery. The more that has been discovered, the more
the mysteries have deepened—and research has unearthed several
orders of magnitude more data on Neanderthals than on any other
hominin than the sapiens. Neanderthals are the first hominin known
to have evolved outside Africa. That they did so under such extraordinary stress, that they were the first of early hominins to be discovered, and that they appear so closely related to the sapiens has
made them rudely allegorical. Their detailed archaeological record,
including burials, set into motion debates that remain remarkably
constant to the present day. They are the archetypal Other of paleoanthropology. They are an enduring enigma in the debate over what,
exactly, constitutes a modern human, and they remain as much a
symbol of the Ice Age as the woolly mammoth.

The Neanderthals' story, one of extreme competition, begins with
their desperate contest against nature in Pleistocene Europe. They succeeded through a blend of genetic adaptations, technological inventions, and social and cultural accommodations. The story then expands
into competitions with other hominins, as Neanderthals pushed outward and found frontiers with the erectines and the early sapiens, and
then as the sapiens pushed back. When they had climate as an ally, at

least during the prevailing glaciations that dominated so much of the Pleistocene, they waxed. When some other factor apparently granted modern humans an adaptive advantage, they waned.

Still, the Neanderthal saga is less about radiations than about refuges. Early hominins wandered the same terrains and obeyed the same rhythms as Irish elks, straight-tusked elephants, and scimitar-toothed tigers and like the great ursids of the epoch, they sought out local environmental sanctuaries while the ice ruled, particularly caves. If the erectines struggled to migrate into glacial climates, the Neanderthals struggled to migrate out of them. They likely evolved in glacial Europe, and they managed to occupy Asia Minor and to push across the mountains of central Asia; there are finds, for example, in the Middle East and Uzbekistan. Instead of bold thrusts outward, however, their long saga is one of a contraction that ended in islands and in caves.

The question of how and why they melted away before *Homo sapiens* is the final great challenge posed by *Homo neanderthalensis*. It involves intellectual competition conducted among contemporary sapiens, because it is a contest played out among those who would understand why the Neanderthals fell to the wayside, and how and why they differed from *Homo sapiens*. In this regard, an assumption has long prevailed that what segregated modern humans from evolutionary ancestors was a soul, or that secularized version of the soul known as reason, or in some other avatar the self-consciousness and moral sense that made for conscience, or in the crabbed jargon of contemporary science, "complex cognitive reasoning and abstraction." The critical fact was not that moderns looked different. It was that they acted differently, and the simplest physical index of such behavior was assumed to be cranial capacity, which is where their reason and moral sense must corporeally reside. Yet much as Linnaeus noted that he could not distinguish anatomically between humans and apes, and had to appeal to intangibles, so T. H. Huxley, pondering the newly excavated skulls of Neanderthals, concluded he could not distinguish between their brain capacity and that of moderns.

Even the intangibles of Neanderthal behavior (including hints at art) blurred the impermeable membrane that putatively segregated the two species. They seemed like two expressions of a common Form. Neanderthals, it seemed, might become yet another of those innumerable footnotes to Plato.

Ice, Fire, and Giants: The European Pleistocene

The mythology of the Norse imagined a cycling of worlds caught between frost and fire that yielded, among its giants and Jotuns, an early human. In the *Gylfaginning*, Snorri Sturluson described how "just as cold arose out of Niflheim, and all terrible things, so also all that looked toward Múspellheim became hot and glowing; but Ginnungagap was as mild as windless air, and when the breath of heat met the rime, so that it melted and dripped, life was quickened . . . and became a man's form." That is not a bad imaginary rendering of Pleistocene Europe, as it swung between glacial and interglacial phases, each cycle emulating a rhythm of extinctions and renewals, out of which emerged a hominin. Especially north of the Mediterranean basin, Europe filled and emptied with strange and outsized creatures.[1]

That Pleistocene past is indeed a foreign country, and it is a place whose conceptualization by observers, like the landscapes they invoke, has kept recycling. The Eddic vision of Snorri Sturluson found itself remade in the Enlightenment into the successive creations imagined by Baron Cuvier when writing during the cataclysms of the French Revolution; the catastrophist vision of Cuvier smoothed and secularized it into uniformitarian spasms in which species and biomes advanced and retreated like ice sheets, an analogy that persisted into the 1940s. Meanwhile, the vision of world cataclysms endures; only the agency of their cause has shifted. The most recent variant has the Holocene as a world whose biotic diversity is being razed and rebuilt and perhaps reconstituted—shaken, not stirred—by the hand of humanity.

Of earth's four great ice sheets, only the Fenno-Scandinavian had contact with Pleistocene hominins. Europe was the scene of repeated glacial advances and retreats as ice spilled outward from the immense mound over the Baltic and glaciers ran down valleys from the Alps, the Pyrenees, the Scottish Highlands, and the Carpathians. Much of the ice-free landscapes between the glaciers remained severely ice-influenced—cold steppes, loess plains, outwash playas. When the ice came it did not merely block biotic colonization but obliterated it, while the lowered seas opened the British Isles to wholesale immigration. Then the process reversed. The ice melted, the land grew to prairie and forest, the rising seas broke Europe into a fractal continent of peninsulas and isles. The life that had fled before the ice now returned. Then the climatic engine again coughed to life; the chill crept back; and Europe became a Nordic Niflheim, a primordial land of mist and frost. Over and again the rhythm replayed: glacial, interglacial, glacial, interglacial—some four major episodes and perhaps fifty minor advances and retreats, a vast syncopation of ice, land, and life. Only the far southern reaches and scattered mountain refugia escaped.

The long-wave frost-thaw cycle cracked and sculpted the biota much as ice broke, scraped, and polished stone. Species adjusted or died out. A revanchist flora returned during each interglacial, and its composition remained relatively consistent through the waxing and waning of the cycles. The fauna, however, showed waves of extinction, and the creatures that disgorged from ark refugia after each ice flood differed dramatically from those that succeeded it.

Europe swarmed with giant mammals, and even as tiny as Europe was—made ecologically smaller by the residual ice and periglacial landscapes—the scene abounded with exotic species. It was as though a woolly Serengeti on steroids had migrated to Pleistocene Europe. The folk legends that ascribed the relic bones found in permafrost or caves to a past time of heroes or to a race of giants were, arguably, not far wrong. The ice and landscape of Europe remained covered with creatures far larger than those that came after the great warmth that ablated away the defining features of the Pleistocene. By the mid-nineteenth century Enlightenment science had begun to substantiate myth. The discovery of Neanderthals founded paleoanthropology much as Heinrich Schliemann's excavations confirming the existence of Homeric Troy established the presence and background of classical archaeology.

The consensus chronology is that around 130,000 years ago northern Europe began emerging out of an intense glacial phase, the Saalian—the most expansive it had experienced, and one that wiped out much of the surface record left by previous, less extensive episodes. The ice was succeeded by an equally intense but much shorter warming spell, the Eemian, that reached a maximum 125,000 years ago, then succumbed to a new onslaught of ice around 115,000 years back. This was the Weichsel glaciation that, after surges and sputterings, plunged to its most ferocious cold snap around 18,000 to 20,000 years ago. A sharp warming began soon afterward that has continued to the present.

During the Eemian, interglacial Europe was warmer, wetter, smaller, and awash with species long since vanished. Hardwoods grew in northern Finland. Southern England had rhinos and hippos, hyenas, bison and aurochs, elk and moose and deer, and elephants. Ice-shorn Europe boasted a menagerie of megafauna, a full ecological complement. There were grazers, browsers, carnivores, omnivores, scavengers; bovids, ursids, elephantids, equids and rhinos, felids, canids. There were lions, leopards, tigers; roe deer, reindeer, horses; wild boars and dholes. When the Weichsel glaciation came they moved or adapted as lands disappeared under ice or became extensive tundra. The survivors included woolly rhinos, mammoths, musk oxen, wolves, and a host of species that found refuge in caves, including the hyenas, lions, bears, and hominins. Every major order was represented, and all seemingly fielded physical giants.[2]

This proved true for hominins as well. For the majority of the Pleistocene, most hominins remained endemic to Africa. One after another they arose; most seeming to mill around the Rift Valley or southern reaches, a few to wander widely. But new species did not seem to result from their far colonizing. Again, over the course of nearly 2.5 million years, *Homo erectus* remained a dominant species. The exception was Africa, where he had originated and where progeny species would evolve from his lineage. Halfway through his life history, perhaps 1.2 million years ago, one such offspring apparently arose, and then began to split again, with some members remaining in Africa and one group trudging northward. As they digressed geographically, they diverged genomically. The northern variant evolved into Neanderthals, the other into modern humans.

The Neanderthals were the hominins of the true Ice Age—not simply the altered climates and landscapes of the Pleistocene but the creatures of a world subjected to the biogeographic frost-thaw cycle of full-blown glaciations in a continent tightly crenulated by seas and mountains and chock a block with cold-adapted megafauna. The Neanderthals were to hominins what cave bears were to browsers and saber-toothed tigers to carnivores. They flourished or perished in similar places of refuge and opportunity.

The Last Human but One

As with all new species, the particulars of Neanderthal emergence are murky. Prevailing research proposes that a million years ago *Homo ergaster*, the African erectine, still dominated the hearth continent while *Homo erectus* had sprawled across the southern tier of Eurasia. Around 200,000 years ago Neanderthals claimed Europe, and *Homo sapiens* was on its way to becoming the dominant hominin in Africa. But how the tribe of hominins got from erectines to Neanderthals and sapiens is unclear.

The usual explanation, as argued by Richard Klein, is that the ergasters began to change around 800,000 years ago. New species appeared in Zambia and Ethiopia, and even 700,000 years ago they were still radiating and migrating through Africa, and then into the Levant and across Europe. The morphologies are unsettled; each discovery seems to spark a new species, complete with a new name: *Homo rhodesiensis*, from the Rift Valley; *Homo antecessor*, from northern Spain; *Homo heidelbergensis*, from Germany and Greece. So say the splitters in systematics. The taxonomic lumpers prefer to place the whole assembly into one plastic species, call them collectively *Homo heidelbergensis* and assume that they are the source for the later hominins. The group that stayed in Europe evolved into Neanderthals. The group that remained in Africa became the sapients. Meanwhile, the erectines roamed Asia and hung on in Africa. The Earth was filling up with hominins as much as with equids and elephantids.

Once in Europe, the ancestral hominin underwent the same evolutionary firing and quenching as other mammals. During full-bore glacials, it found itself pushed to the margins, and even outside Europe proper, into the Levant, perhaps North Africa, and across the Trans-Asian mountains. During interglacials, it was pulled back. This climatic bellowing continued for several hundred thousand years, during which the "archaic" hominin became a Neanderthal. By 400,000 years ago, early Neanderthal features were in evidence; by two hundred thousand, most were present; and by one hundred thousand, "classic" Neanderthal anatomy was the norm. The Neanderthals, that is, evolved at the same time as modern sapients; they just did it in a different place. Uniquely, among hominins, they emerged amid cold and mountains.

Their morphology reflected those circumstances. They were big, probably a reflection of Bergmann's law that relates a species' bulkiness to cold climates. But their bulk was composed of sturdy bones and large muscles, not blubber. They were predators: an apex carnivore that lived by hunting and pitted bone, muscle, and stout spears against the other giants of Pleistocene Europe in close combat. They hunted bison, horses, assorted bovids (from red deer to reindeer), and no doubt the occasional mammoth. They suffered the pains and pathologies of such a life. Fossil skeletons show abundant broken bones and dislocated joints, symptoms that likely indicate frequently torn ligaments and tendons as well, with an average life expectancy of thirty-five to forty years. They suffered from arthritis. Their teeth wore down from heavy use and were pitted with dental caries.

They had something else as well, which was connected with those brains. They had the genetic and anatomical makeup for speech, including ample brain mass in the parietal lobe, which is responsible for language, learning, and memory. They could manufacture complex tools, from Mousterian flakes to spears to toothpicks. They could communicate and organize socially. In fact, they had a social order that included caring for the sick and injured. They lived according to learned behavior and according to what is hard not to call a moral sense, in that it involved deliberate choice and surely forms of consciousness.[3]

For over 200,000 years they flourished under the most extraordinary circumstances. And then, having survived two major glacial cycles, they expired. Before the Weichsel glaciation reached its cold climax, they were gone. Between 30,000 and 35,000 years ago, they disappeared, among the earliest of the megafaunal extinctions that closed the Pleistocene. The last known holdout was a cave-pocked cliff in Gibraltar, where the final ember of a unique cold-world species is found in the ashy remains of a hearth.

The mysterious demise of the last Neanderthal thus pairs with the mysterious origin of the first, and together they frame not so much the evolutionary span of a species as the life cycle of an idea and a discourse. From their initial discovery, Neanderthals have fascinated as a prehistoric Other, and what animates research is the relationship between them and the sapiens, *Homo sapiens.*

Other early hominins could be differentiated by obvious morphological differences in thumbs, bipedal adaptations, molars, and mandibles. The last characteristic to appear—the one that most pointed to modernity, it was assumed—was a large brain, which was needed to house that unique spark of consciousness and conscience that best characterized humanity. But as Huxley early observed, it is possible to select a series of recent human skulls that can "lead by insensible gradations from the Neanderthal skull up to the most ordinary forms."[4] If brain size was not diagnostic, then perhaps the stuff in brains—that is, culture—was. But does not culture, too, have its infinite gradations and missing links that make the clear emergence of a new expression difficult to discriminate? All this is a problem in the phylogeny of ideas. An answer depends on what comparisons and criteria are chosen.

If the initial debates over Neanderthals introduced the scenario for discovering early hominins, they also laid down the terms of discourse for debating what those relics and artifacts mean. The Neanderthal story challenged the typical formulas. The usual criteria do not seem to work, although such failures may reflect comparisons that are unsuited for the task; the narcissism of small differences may inflate what is trivial and overlook what is obvious. Another comparison might allow an alternative, for Neanderthals were, after all, only one among many megafauna mammals, nearly all of which vanished.

The Cave Bear Clan

The Neanderthals' nearest megafaunal cognates, and rivals, were the ursids. Bears were big; they were adapted to Pleistocene cold; they were omnivores; they shared similar habitats and phylogenies; and they have come to occupy a comparable slot in the moral geography of the Pleistocene. Moreover, as with Neanderthals (and for similar reasons) their fossil record, as Björn Kurtén notes, is "excellent," as complete as that for the horse, and "as regards the Pleistocene bears of Europe, almost incomparable." The cave bear, in particular, presents an eerie echo of hominin history and discovery.[5]

Ursid phylogeny followed the epoch's classic scenarios. When the Pleistocene opened there was one species, *Ursus minimus*, which was evolving into *Ursus etruscus*, and then they radiated into cave bears, black bears, brown bears, and polar bears. Both black bears and brown bears spread throughout Eurasia, with the black bear speciating into Himalayan and Malayan variants. Both black and brown crossed into the Americas, the brown bear (*Ursus arctos*) very late. The polar bear (*Ursus maritimus*) evolved during the last glaciation and adapted to life on ice. Throughout, the European lineage delivered a sequence of caving bears roughly corresponding with the cadence of glacials and ending with the true cave bear, *Ursus spelaeus*, which became extinct at the same time as the Neanderthals, with another ursid replacing it.[6]

The cave bear thus stands to ursids as mammoths do to elephantids. It was the culmination of the Pleistocene radiation: the "most bearish of bears." It was huge and powerful, and like mammoths apparently an opportunistic omnivore that mostly consumed plants, which accounted for its colossal head, which served to anchor the massive musculature needed to work its powerful molars. Like all ursids it sought dens for birthing and hibernation, which led it to caves. There it left an unrivaled record. Europe's caves have excavated the remains of between thirty thousand and fifty thousand bears. So dense is the bony residue that its decomposition left deep deposits of phosphate, whose quarrying is what first brought the relict bear bones to the attention of naturalists and intellectuals.[7]

The cave bear was one of Pleistocene Europe's endemics. It arose as a unique outcome of glacial cycling acting on Europe's terrains and their resident fauna, shaped, that is, by the rhythms of ice and sun and a mosaicked landscape of biomes and caves. Historically, the cave bear appears during the last two waves of glaciation and disappears with the onset of the Holocene warming. Geographically, it ranged from Europe south of the ice sheets and eastward across the Transcaucasus Mountains, with evidence it might have reached Morocco and the Trans-Asian ranges farther east. Ecologically, its population was likely limited since they were, individually, immense and territorial omnivores, largely browsers, and so could not have thrived in vast numbers. They knew few predators, the wolf probably being the most prominent. Mostly, they died from natural causes. Diseases were common; teeth were worn, broken, even corrupted by caries; bones suffered from infections and mechanical damage. Not a few afflictions resulted from the burden of bearing its sheer bulk and strength. Like all bears it sought dens for birthing and hibernation, which led it to caves. There it left an unrivaled record. Europe's caves have excavated the remains of between thirty thousand and fifty thousand bears (some observers believe that most of the cave bears who have ever lived may be among that extant record).

All this makes the cave bear a model species for the European Pleistocene. But what enhances biological interest is its value as an analog for Europe's endemic hominin, the Neanderthal. In its evolutionary history, in its geographic range, in its distinctive pathologies, in the manner of its discovery, in the extraordinary richness of its paleontological record, the cave bear is virtually interchangeable with Neanderthals, and this equivalency apparently extends even to the manner and timing of its extinction, as it retreated to more remote sites and refugia, its marvelous adaptations no longer adequate, its niche challenged by an aggressive and omnivorous rival. For *Ursus spelaeus* that competitor was the brown bear; for embattled Neanderthals, *Homo sapiens*. In truth, how both European endemics related to the sapiens is what has moved inquiry from a passing curiosity to what might be characterized as a cultural compulsion.

Eventually the cave bear died out; isolated into remote sites and refugia, its marvelous adaptations no longer adequate, its niche challenged by another ursid—the aggressive and omnivorous brown bear. In this way the ursid story parallels, once again, the hominin story, as sapiens moved in from the periphery of Europe to replace the endemic and embattled Neanderthals.

Like Neanderthals the bear clan looms large in contemporary imagination, both for their emblematic standing in environmental ethics and for their intellectual potential as a model species for extinction. Like elephants, bears have too many features that resemble humans' and occupy common, even shared settings. They pose questions about how similar species compete and die out.

The brown (grizzly) bear has evolved into one of a handful of canonical symbols of the wild. In Europe perhaps only the auroch and tarpan, prehistoric bulls and horses, claim comparable status. In North America its primary competitor as poster child of the wild is the wolf. And of course there is its white twin, the polar bear, the ursid most specifically adapted to Pleistocene ice and accordingly vulnerable to the collapse of the Arctic's frozen sea; it stands for the shrinking habitat of Pleistocene ice as the California condor does to land. In this way, contemporary environmentalism has again called upon an emissary from the Pleistocene to serve as a symbol of human malfeasance, a living metric for humanity's mismanagement of Earth.

But humanity's continued engagement with ursids is no less vibrant intellectually. It has to do with how we might understand the relationship between the two genera and how one might serve as a surrogate for the other. The brown bears and the sapients bore down not only on each other but on their genomic relatives; each was an aggressive omnivore and a creature with few natural enemies save each other—and of course their own kind. The cave and brown bears might well serve as model species for understanding the dynamics between Neanderthals and sapients. The dramatic plot of the last cave bear might complement the sad saga of the last Neanderthal.

Their differences, however, complicate simplistic analogies, and the ursids again show how. The customary characterization of Neanderthals points to their obvious somatic suitability to thrive amid glacial cold and its creatures. The argument is, Neanderthals represent an anatomical adaptation rather than a cultural one; this is what truly distinguishes them from the sapients. Yet this thesis—that anatomy is destiny—is challenged by a genera of Pleistocene bears from the New World. *Tremarctos* ranged from the Andes to the southern United States. In the latter it evolved into an "uncanny" resemblance to the European cave bear "in some of its most conspicuous features," even earning the title "Florida cave bear." This expression of evolutionary convergence suggested to Björn Kurtén that "special adaptation to glacial conditions can have played only a minor role in the evolution of the distinctive characteristics of the cave bear." Rather, he thought, what pushed the two bears into convergence was a shared drive

toward a vegetarian diet and a reliance on size and strength for protection.[8]

Neanderthals were certainly adapted to Pleistocene Europe, but
how, and in what senses, and in what ways were those adaptations
mostly morphological? It begs the question: why did the Neanderthals "fail," since the cave-dwelling hominin, like the cave bear, faltered *during* the last glaciation, not afterward? The speculation turns
on those big heads, the one specialized to anchor muscles for chewing
and the other to hold an oversized brain. The argument that Neanderthals had some culture—could talk, think, make art—won't go away.
On it pivots much of the moral drama of their disappearance. If they
resembled cave bears they vanished because the collapse of the last
glacial left them morphologically maladapted to a warmed world. If
they resembled grizzlies (or their hominin cognate, the sapiens) they
perhaps had the capacity to adapt behaviorally but vanished because
of sibling rivalry, both overt and covert, among the hominins. Among
the causes, how much was nature and how much was culture?[9]

Knowing how the Neanderthals ended helps sharpen the question
of how *Homo* began. Neanderthals are one of the great counterfactuals
of paleoanthropology, and they seemingly provide an aesthetic and
ethical closure to the narrative started with the habilines. An answer to
their demise, if it comes, will likely be heard from the caves that are for
both ursids and hominins the great archives of the Pleistocene.

The Last Made First

The Neanderthals were the first and in many ways the last. They
were the first pre-sapient to be discovered, and hence kindled the
modern debate about human origins. They were the last of modern
humanity's predecessors, and thus sparked a long discourse about
how and why they disappeared. The purpose of both discussions was
of course to sharpen understanding about what makes humans human
and moderns modern.

That query takes several forms, but they all hinge on the question
of survival, and survival segues into issues of competition. Hominins

competed with other megafauna, Europe's Pleistocene hominins competed with other hominins, and the discoverers of Neanderthals competed among themselves for explanations. Each struggled for existence according to its own terms. Among megafauna the terms were the standard measures of evolutionary success. Among hominins it involved some new factor, unique to modern humans, that seemed to intrude to tilt the scales between those who flourished and those who faltered. Among theories, the process was one in which methods and criteria came and went while the core questions remained unanswered, because, other than their details, they were unanswerable by such processes alone.

Models of Migration

In recent years the geographic and genomic data have converged to tease out the particulars of this complex narrative. Fossils have identified the locations while genes have isolated the story of colonizing, competing, and hybridizing, and the tracks of other species have become firm enough to serve as possible analogues for hominins. To a great extent those conceptualizations hinge on the role of refugia—safe havens that served both as a sanctuary during glacials and as a source of repopulation during interglacials. They were also a Pleistocene paradox, since refugia, like caves and other redoubts, could be a trap as well as a preserve.

Europe's southern rim of mountains and, along its eastern margin, the Carpathians formed the frontier between ice ages and sanctuaries. To the north, and along mountain summits, ice and tundra prevailed, along with the species adapted to them. To the south, temperate species could survive. The peninsulas—Iberia, Italy, Greece, and, marginally, Turkey—were like holding pens into which the glacial surges drove species, with the mountain ranges shutting the gates. Here they milled, mingled, and sifted themselves until a warming pried the gates open again, and they stormed out. Greece was doubly shuttered by its own mountains and by the encasing Balkans, making escape even more formidable. Not all species made it into the pens, and not all in the holding pens were able to reclaim their old estates once released.[10]

The locations of refugia were fixed by geography; what varied was the intensity of glacial surging and collapse. As the cold intensified, the sanctuaries shrank and pared the population of survivors by testing their capacity to resist and endure. But when the ice melted away, the refugees differed in their capacity to seize the new opportunities being presented. They needed less the ability to hunker down than to roam. The early interglacials were land rushes in which species differentially staked their claims, in what Godfrey Hewitt describes as "paradigms" of recolonization. These are the small suite of models that depict characteristic patterns of biotic advance and retreat. The grasshopper model, based on the meadow grasshopper (*Chorthippus parallelus*), spilled out of the Balkans and Carpathians and swarmed throughout Europe. So rapidly did the species spread that competing organisms could not breach the Pyrenees and Alps in time and were held within their pens. The hedgehog model (*Erinaceus* spp.) involved multiple streams of émigrés, one from each of the refugia, that collectively reclaimed lands north into Scandinavia. The brown bear model (*Ursus arctos*) describes a pincer movement, with a western thrust out of Iberia and an eastern one out of the Carpathians, that met along a muddled frontier. The speed of one species could effectively deny slower colonizers the opportunity to escape.[11]

Which model did Europe's Neanderthals follow? Obviously those hominins that were present joined the throngs that shoved and strode between perimeter and core. Neanderthals had participated in this cycling for at least half the Pleistocene, and they had persisted through at least two major glaciations. If they followed the best-guess tracks of modern humans, they resembled grasshoppers in their advance and perhaps bears in their retreat. In deglaciated times they spread quickly across plains, then seeped up the slopes and into mountain valleys. In glacial outbreaks they retreated across the mountain frontiers and sought sanctuary in protected sites and caves. Over and again, they likely repeated this scenario.[12]

Then, prior to the last glacial maximum (16,000 to 23,000 years ago), they vanished. It is not simply that they did not survive that final brutal crush of ice. They faltered before it arrived. Most disappeared between 30,000 and 35,000 years ago. "We will never know for certain what happened," observes archaeologist Ian Tattarsall. "All we can say with assurance is that in the end, the moderns won out." The last known holdouts are in southern Spain along the far fringe of the refugia.[13]

Exemplars of Extinction

During the Pleistocene species had ebbed and flowed into and out of refugia many times, but the last cycle was different. It particularly targeted larger fauna (around one hundred pounds). But unlike past episodes, those fauna were not replaced; some regions were much harder hit than others, and this time the rhythm of extinction veered more and more from the chronology of climate. Some new presence apparently entered into the melee. Here the paleo record on hominins and the rest of creation converges. The distortions coincide with the arrival of modern humans.[14]

The losses were extensive. The waves of extinctions were present even when the onset of the Pleistocene was defined as 1.8 million years ago. Of some 514 mammalian genera identified from the fossil record, 207 (or 40 percent) disappeared. The bigger the beast, the more likely and harder its fall; the farther from Africa, the more widespread the extermination wave; the more recent the entry of humans, the more sudden the die-off. Over and again the unifying factor in whether big beasts died out or endured seems to be their relationship to the sapients.[15]

Three scenarios seem to apply when discussing Pleistocene people and megafauna. In the first, humans cause the beasts to expire. The megafauna are hunted or hounded into oblivion or have their habitat so overturned that they cannot survive. Humans may prey overtly on them, or they may simply help catalyze conditions that make survival impossible, for it is not necessary for humans to kill off these creatures by contact but to simply tip the scales in ways that favor hominins and disadvantage others. This—humans as exterminators, whether by direct or indirect means—has been the classic explanation. It hinges on the perception that modern humans are intrinsically destabilizing, or even endowed with an ecological version of original sin. Species thrive or collapse depending on the conduct of people, whether by malice, whim, or clumsiness.

Of course, some species would have died out without humans present, but the record does reveal an eerie coincidence of human arrivals and megafaunal departures. The later and smaller the land colonized, the more severe the outcome. Africa suffered the least; only some 14.3 percent of its megafauna expired. Australia experienced the extinction of 86.4 percent. North America lost 73.3 percent, and South America 79.6 percent. In historic times Madagascar, a microcontinent, witnessed an almost total extinction; so did New Zealand, the Antilles, and the larger islands of the Mediterranean. The wave rolled over new lands more than old, and small lands more thoroughly than large, and its timing corresponds far better with human colonization than with climate. The African exception is understood to result from a lengthy period of coevolution in which modern humans were not a new factor but one that emerged, with tedious patience, over geoevolutionary time, and hence mutual accommodations. Interestingly, in recent centuries the colonization of Africa from outside did threaten to wipe out its large mammals.[16]

The other two scenarios for how the Pleistocene's people and its megafauna interacted play out after the Pleistocene has passed. In one, select fauna become domesticated. In effect, they join the human community, or even become family members (in ancient times they sometimes underwent adoption ceremonies). Domestication thus modulates both the geographic and genomic narratives by having people relocate the habitats of those species, and even substitute cultural selection for natural selection. So powerful is the process that the onset of domestication is often taken as a marker for the Pleistocene's end. The last scenario is more recent yet. It describes the practice of deliberately protecting species by establishing reserves or reintroducing species that have been lost.

Thus, for elephantids, the mastodonts died off, very possibly with active hunting by hominins; the Asian elephant survived by being domesticated; and the African elephant persists through active conservation measures. The equids also died off, save for the Przewalski horse from central Asia, which was domesticated and then shipped around the world to repopulate the lost species, and its descendants continue today as ceremonial creatures or family pets, while the Przewalski horse finds itself the object of reintroduction and special protection. Among the ursids, the cave bear seems to have passed away from natural causes while the brown bear was hunted to extinction in many places; in more recent years the latter has enjoyed legislative protection.

What they all share, however, is that control over their evolutionary future has passed from the forces that began the Pleistocene to those that ended it. Clearly, in its early parts the epoch underwent plenty of extinctions for which hominins were at best a negligible participant, and in fact a victim. After the epoch ended, there are, with equal clarity, ample examples of humanity driving species into oblivion. When, exactly, the transition between them occurs is unsettled—probably sometime during the last glacial cycling. But the relationship between hominins and comparable megafauna has become another indicator of what makes modern sapiens what they are.

A similar scenario applies to hominins as well, for they also experienced cycles of radiation and extinction. Like equids and ursids, they underwent an expansion in the early Pleistocene and a severe contraction in the late. And like the others, they thrived or fell at the end of the epoch based on their relationship to modern sapiens.

How did those hominins come and go? One explanation of course is that they emerged and went extinct in the same way other megafauna did. They obeyed the same implacable cadences of natural selection, that evolutionary fugue between genomes and geography, as fully as ursids and equids. This scenario surely dominated in the early phases of the Pleistocene. But another possible suite of narratives exists that derives from historic, post-Pleistocene times. The analogies here are ones in which newly arriving humans have broken or wiped out indigenous peoples, sometimes by outright conquest or enslavement, sometimes by introducing diseases, sometimes by so upending habitats or social orders that survival of the indigenes and their folkways is impossible. Sometimes, too, the drama has been one of absorption, a process between hybridization and domestication.

The most recent, best-documented contact stories involve the era of European imperialism. The melancholy collapse of indigenes (and of indigenous biotas) impressed mightily the Enlightenment's corps of exploring naturalists, including the codiscoverers of evolution by natural selection. But the narrative of decline and fall became a staple trope and is one that contemporary historians have continued. Alfred Crosby gave it an environmental coloration when he made the native Guanches of the Canary Islands, first resisting the invaders, then collapsing through disease and war, the model for contact between an expanding Europe and indigenes everywhere. The Guanches were, "with the possible exception of the Arawaks of the West Indies, the first people to be driven over the cliff of extinction by modern imperialism." Here is a model that might also describe the encounter between expanding sapients and indigenous bison, giant wombats, and dodos.[17]

Those who study natural history are likely to portray the demographic collapse of Pleistocene hominins by analogy to the implosion of cave bears and mammoths. Those hominins were simply part of the menagerie, subject to the same rules of survival. Those who study human history are more likely to project back onto mastodons and ground sloths the record of modern sapiens who eliminated species and upended habitats as they colonized new lands. In addition to statistics they can offer figures of speech, in particular the literary conceit of the sole survivor of a vanishing tribe. In this sense, Ian Tattersall's *The Last Neanderthal* is a descendant of James Fenimore Cooper's *The Last of the Mohicans*, as the Last Neanderthal might also stand for the first Guanche.

Both interpretations pivot on when humans became a force of

nature that could presume to compete with natural selection, and they focus on the demise of the Neanderthals because their extinction seems to have been out of sync with the climate. "Their abrupt demise," concludes Ian Tattersall, "must thus have been due to an entirely new factor"—"almost certainly"—the sapiens. The cultural fascination, however, is less about the Neanderthals than the sapiens, because the issue is not whether sapiens could wipe out species and overturn environments in principle; there is abundant historic evidence that they could. The issue is what happened in practice because knowing how it happened describes the choices the founding sapiens made, which defines their character. A chronicle can correlate Neanderthals with climate and other creatures. A narrative will turn on moral judgments that bones and genomes cannot by themselves decode.[18]

Founding Fathers

The original discussion of the Neanderthal evidence was premature—the bones were distorted by disease and the announcement was skewed by timing. When Professor Hermann Schaaffhausen presented the bones to the Lower Rhine Medical and Natural History Society in 1857, some doubted that the bones were human or, if human, asserted that they were the remains of a degenerate. Schaaffhausen thought they likely represented a member of the barbarian tribes that Rome had sought to pacify. Later, Professor Franz Mayer thought them the bones of a Cossack deserter from the Russian army that had chased Napoleon back to France. No consensus existed because there were no criteria by which to evaluate and no conceptual matrix by which to derive criteria. All that changed with Darwin's *Origin of Species*.[19]

The Descent of Man

The question of whether organic evolution ought to apply to humans was unavoidable. Soon after the *Origin*'s publication, Charles Lyell revisited collections exhumed several decades previously from the Engis

Cave near Liège; these included hominin bones that now assumed
new significance. Meanwhile, George Busk translated and published
Schaaffhausen's paper in 1861 and sent T. H. Huxley casts of the
bones, which assumed the status of empirical tests of Darwin's theory, and evidence on its behalf. They required only interpretation,
which depended on context. In short order all the great promoters of
organic evolution—Charles Lyell, Thomas Huxley, Alfred Wallace—
weighed in to establish that context. Lyell compared the bones to
fossils, Huxley compared the bones to other bones, and Wallace
compared the bones to behavior. Collectively, they bonded the Neanderthal skeletons to the modern disciplines of geology, biology, and
anthropology.

For Lyell context meant that the hominin bones had to lie in the
same strata as the fossils of extinct creatures; this would date them
and render them part of the same processes of evolution and geologic
deposition as other bones in the mix; they were themselves index fossils. The dates were vital, because if the cranium were recent, it was
likely the remains belonged to a modern "degenerate," but if "very
ancient" then they bespoke "a less advanced stage of progressive development and improvement." Whether it was plausible to place humans within Darwinian evolution depended on the time allotted and
their shared occupancy with other vanished creatures. Did that mean
it had actually happened? Lyell finessed with a double negative: "We
certainly cannot escape from such a conclusion"—that "man himself
has been derived by an unbroken line of descent from some one of
the inferior animals"—without abandoning Darwinian evolution.
Lyell softened that conclusion, however, by invoking "the improvable
reason of Man" and the "picture of the ever-increasing dominion of
mind over matter."[20]

For Huxley the issue was morphological, a question of comparative anatomy. He delivered a series of lectures, published in January
1863 as *Man's Place in Nature*. The Engis skull was a crisper specimen
than Neander, but neither, Huxley declaimed, could "be regarded as
the remains of a human being intermediate between Men and Apes."
By now George Busk had received the Gibraltar skull, and in January

1864 William King, professor of geology at Queens College, Galway, decided that the Neander skull belonged to a new species, closer to apes than to modern humans, which he named *Homo neanderthalis*. Huxley immediately denounced that conclusion and argued that the range of variability within modern skulls so overlapped with that of the few Neanderthal skulls that there was no clear distinction sufficient to warrant taxonomic splitting. There were features that differed—some "extraordinary"—but they were variants from moderns, not links on a chain to apes.[21]

So Neanderthals were not the *experimentum crucis* that would demonstrate descent by evolution from an ancestral primate. As more bones accumulated it became clear that they differed from modern humans and might constitute a separate species; but if separate, on what basis could that judgment be made? The evidence, it seemed, could not come from the fossils alone but would require some other measure. Here, what matter couldn't solve, perhaps mind could. Previously commentators had only to point to brain mass as conclusive. With Neanderthals that was no longer possible. Their brains were plenty big; it was how it worked that mattered. Brain had to be converted into mind.

The Descent of Discourse

One of the great trekking naturalists of his time, Alfred Russel Wallace had independently discovered evolution by natural selection and appreciated that theory as few could. He also had great sympathy (as Darwin did not) for the "primitive" peoples of the world and wanted to argue not only for the descent of humanity from an ancestral species but for humanity's essential unity and uniqueness. Humanity was "not only the head and culminating point of the grand series of organic nature," he insisted, but "in some degree a new and distinct order of being." The measure of this distinctiveness was not the brain but what went on inside it. What set modern humans apart was "their superior sympathetic and moral feelings," which fitted them for a social order that could stand between them and brute nature.[22]

For Wallace, humanity—that is, the sapients—was "a being apart," in "some degree superior to nature" because it had transcended natural selection and "was no longer necessarily subject to change with the changing universe." With superior intellect, humanity could provide for itself what previously could only come from nature by adaptation; with superior "sympathetic and moral feelings" humanity became a social creature who changed the terms of survival such that the ill, the poor, the weak, and the helpless are not cast aside but assisted. Instead, the functions that used to reside in nature have come to reside in humanity, such that evolution no longer proceeds, for humans, by changes in anatomy according to the utilitarian calculus of natural selection but through changes in mental and moral states. In contemporary phrasing, cultural evolution replaced organic evolution.[23]

While this process removed humanity from the rude nature of the lower animals, it also joined humans among themselves, because it meant physical characteristics such as those outward traits that distinguished the various races were not definitive; "mental" traits had replaced them. The tricky part was identifying the transitional phase between the time when natural selection did operate and had produced racial features and the time when mind and morals had taken over. When did it happen? And how could such gradual change be identified in the geological record?

While Wallace did not speak directly about the Neanderthals, he did address the question of early humans. He observed, after Richard Owen, that anatomy alone made it difficult to discriminate between apes and humans, so he offered a surrogate divide in the form of an "intellectual chasm." This strategy, however, only replaced a tangible trait with an intangible one, proposing that behavioral differences might substitute for anatomical ones. Most of his contemporaries would have assumed that morphology matched mind, that a big brain was a suitable measure of behavior, which is to say, of character. But if, as Huxley indicated, crania were inadequate as criteria, then something else would have to serve.[24]

This reasoning placed the burden of proof onto behavior as an index of culture and onto artifacts as an index of behavior. The putative culture of Neanderthals, that is, could only be inferred from other evidence. The more worrisome issue, however, was that analysis yielded the same kind of gradations that had compromised the study of Neanderthal skulls. Thanks to Wallace those gradations now resided in behavior as well as anatomy. The human essence was reembodied into anthropology, the soul reincarnated into culture, and moral philosophy not only secularized but scientized, as it were. What endured was the perceived need for humanity to be different and to find in that difference some basis for explaining the why and how of human existence.

Further discoveries seemingly made that task harder rather than easier. In order to be understood they required context; for context, they required comparisons and contrasts; for comparison and contrast, they required firm criteria. But substituting culture for nature only replaced one variable with another. However necessary, a distinction was arbitrary, for while it made comparison possible, it had no objective basis, but was a choice made according to the values of those doing the choosing. Still, to the extent that behavior was the measure of modernity, it became essential to deny Neanderthals any expression of art, higher reasoning, or abstract logic that would edge their behavior across that tipping point.

Clearly Neanderthals were different, but their place among the evolutionary phylogeny of hominins may matter less than their role in the evolution of ideas. It seemed necessary to split off ancients in order to define moderns. It was possible to place erectines easily, if roughly, into a historical chain, but Neanderthals were not ancestors so much as cousins, and they practically begged to be different. Yet the basis for that discrimination was unstable, because it relied on a definition that moved from brain size to mental capacity, and this was itself a construction of the mind, and worse, of the mind contemplating the mind.

The real challenge, however, came not from the Neanderthals, who were at least familiar—the model species, as it were, for the study of human evolution. Their culture was as incrementally distinguished from moderns as their anatomy. The contrast between the two species was in some ways statistical, like identifying a mean tide.

The Neanderthals had a big brain in a big body, both different from sapiens but clearly analogous. They made a convenient, perhaps essential, Other, part of a complicated duality that allowed for comparisons. The more serious test would come from a species whose morphology was unblinkingly different but whose behavior was likely similar. It would come with the discovery of a hominin that had a proportionally big brain in a small body that seemed to show traits of behavior otherwise claimed by the sapiens.

Places of Refuge, Sites of Revelation

Caves were domiciles, refuges and sanctuaries, oracles, and entries to other worlds. They collected the outside world, sieved its abundance, and stockpiled the survivors. Cave bears went to caves to hibernate and die, Neanderthals and sapiens to live and leave relics. The known world of Pleistocene hominins is largely the microcosm preserved in its caves. They are, in truth, immense caches of data. Their microgeomorphology documents changes in climate and terrain; their biotic residues, rhythms of biogeography; their stratigraphy, a chronicle of natural history; their bones, a depository of relics, curiosities, and cultural histories.

Each data hoard requires exegesis, each has its own instruments or methodology, and each exhibits traditions of interpretation. It is all possible because caves collect and concentrate. Without them information about the Pleistocene and its hominins would blow in the wind like loess and wash away into obscure deltas and placers of former lakes and streams. For bears the deposition was the inevitable result of denning and hibernating. They died where they lay, leaving behind bones and phosphate like etched runes. For various predators, like lions and hyenas, the caves held whatever they dragged within. For Neanderthals, the cave was a shelter, at least seasonally; and some of what they left was simply refuse; and some, what they chose to bury. Among the last were other Neanderthals. Only hominins deliberately interred remains as a record, and only hominins would deliberately exhume them to read their odd script.

There are certain symmetries in the long epic of the Neanderthals, not only in their phylogenetic life cycle, but in the story by which that saga became known. Both story and fossil begin and end at Gibraltar.

The first known Neanderthal skull came from the island's Cochrane's Cave in 1848. It was housed as an unlabeled curiosity in the museum of the Gibraltar Scientific Society. The society went defunct in 1853. Four years later, in 1857, the German Neander Valley skull was announced; in 1859, Darwin published the *Origin;* and in 1861 George Busk, using a cast of the Neander skull, translated the discovery for *Natural History Review* and sent the cast to Huxley for comments. Two years later Busk visited Gibraltar, and later that year he was sent the odd skull from the museum's collections, which were then being sifted through, and he immediately identified the "pithecoid priscan" man's skull as identical "in all essential particulars" to "the far famed Neanderthal skull" and concluded that it could not be the eccentric token of an individual but the remains of a "race" that had extended "from the Rhine to the Pillars of Hercules." One odd skull was an anecdote; two, a fact. The Gibraltar skull confirmed Neanderthal as a species of early hominin.[25]

That the evidence should come from caves, however, seems wholly appropriate, since caves were refugia that protected Neanderthals when they lived and have shielded the record of their past. The relics of the last known Neanderthals come from the Zafarraya Cave in Granada and various caves on Gibraltar, particularly Gorham's Cave, where hearth stacks upon hearth, between 24,000 and 28,000 years ago. The last Neanderthals apparently coexisted for several thousand years with sapiens—on what terms is unclear. But then, the terms by which moderns understand Neanderthals may not be known with any surer clarity.[26]

That the caves should be sited on Gibraltar seems no less appropriate, for the Pillars of Hercules were considered by the ancients as the practical limits of the known world and by early moderns as the symbol of the limits of ancient learning. To pass beyond the Pillars was to go from past to present.

Thomas Huxley once noted that "[a]ncient traditions, when tested by the severe processes of modern investigation, commonly enough fade away into mere dreams: but it is singular how often the dream turns out to have been a half-waking one, presaging a reality." By such means, a connection to antiquity has not utterly lapsed, and its relics remain buried in the civilization's intellectual sediments. Like Neanderthals, they remind us of an uncertain past and the ways it continues to intrude on the present. The cave is the connection between them.[27]

Long before the advent of modern science, caves were considered portals to other kinds of knowledge. They were the gateways to another world—a world at once deeper and more foundational, and a realm of the spirit. They were, for epic heroes, the entry to the underworld. Here Odysseus and Aeneas passed into the nether realm to acquire special knowledge, a journey the Renaissance would revive, along with the recovery of ancient texts, with Dante. Prophets, sibyls, and oracles guarded the entries to such sites, for here communication with the gods was possible. When Aeneas reaches the portal the prophetess cries out, "Time now to ask your fates." In a more secular and skeptical world, it is still where communication with the past becomes possible.

So it seems especially appropriate that the heroes of early paleontology passed into caves, as Thomas Grebner, professor of theology and philosophy at the University of Würzburg, did the Zoolith Cave of Gailenreuth in 1748. He described his "descent" with classical allusion by invoking the Muse, entered "a cleft in the woods, in a desolate, distant area," and, finally, his tortuous passage lit by torches, coming upon the wreckage of skeletons and "bodies turned into stone." To quote: "The gruesome, terror-inspiring, immeasurable cavern, the lightless grotto, O Muse, bring back to my mind, and those death-like realms!" The account comes from Homer by way of Virgil through Dante. The dead will speak, even if they do so in allusions and often in riddles. The words that emerged from that peculiar communion would demand parsing and commentary.[28]

The ultimate oracle was of course that at Delphi, on the slopes of

Mount Parnassus. Here the voice that emerged from the cave had a mystical aura, probably stimulated by inhaling a trance-inducing *pneuma* wafting through the cleft. The utterances were sometimes specific, but often cryptic and allegorical. A coterie of priests and acolytes had to interpret the result, and if the prophecy was wrong, the error was presumed to lie not with the utterance but with the glosses. Much the same dynamic continues, with professional commentators interpreting the ambiguous utterances that arise out of the conceptual pneuma of paleoanthropological discovery.[29]

The journey to an oracle to hear the secret wisdom by which to explain who we are and what we should do endures. The oracle's message, however, resides neither in reason nor in revelation alone, however evolved and reborn over the centuries. Instead it harks back to the inscription that hovers above the site at Delphi: "Know thyself." It is a difficult commission, which moderns attempt to solve by asking one another, and by seeing themselves in others, among whom are the Neanderthals.

So seekers still enter the caves to speak with the dead, and the Neanderthals—or their ciphers—continue to commune, but they do not always speak a language moderns understand. So the interpretation goes on, as spectrometers and microwaves replace the pneuma, and as bones call out in Delphic vagueness, and as taphonomists and DNA decoders try to parse the ineffable into the cogent and rephrase prophecy into hypothesis. Over the years the range of uncertainties has narrowed, such that the utterances from the bones are clearer and less clouded with babble, but the uncertainties ultimately persist, because they cannot be answered by such methods. At issue is not so much a question as a quest, perhaps the oldest.

Dwarves on the Shoulders of Gyants

In 2004, a team of Australian and Indonesian archaeologists announced the discovery of new hominin bones from the island of Flores, in the Lesser Sundas. They found the nearly entire skeleton of an adult female who had stood a meter tall, weighed thirty-five to sixty-four pounds, and died about 18,000 years ago. Its cranium was the size of a chimpanzee's; its bipedal features were those of an australopithecine; the shape of the skull was between an erectine and a sapient; and the associated technology, including fire, spearheads, and group hunting were complex. As Marta Lahr and Robert Foley put it, "The find" is that "of a pygmy-sized, small-brained hominin, which lived as recently as 18,000 years ago, and which was found on the island of Flores together with stone tools, dwarf elephants and Komodo dragons. Discoveries don't get better than that."[30]

Because of its scale—and the popularity of the *Lord of the Rings* movie trilogy completed the year before—the team nicknamed the specimen the Hobbit. It resembled nothing in known hominin phylogeny and suggested a new species, and one whose difference did not depend on the possession of one or another more advanced trait but in how the various traits that make up the evolutionary heritage of modern hominins come together. The Flores find is, as two commentators put it, "the most extreme hominin ever discovered."[31]

On the Edge of Creation

Homo floresiensis was a startling revelation. It challenged the simple Out of Africa narrative in which hominins had originated only in Africa by demonstrating in situ speciation even beyond the shores of Asia. It delaminated the prevailing phylogenies by which traits that culminated in modern humans had appeared in rough sequence, since the features of the new hominin displayed separate staging of such features as cranial size or arm length. It challenged the discourse about brain and mind in confirming that modern behavior might not align with big brains. It upset the comfortable dialectic—almost a ritual discourse—that had evolved between Neanderthals and sapients and that had for so long shaped discussions about modernity. It was not a missing link between known species so much as a lost link or an extra link on a parallel chain. Before Flores, as one discoverer, Peter Brown, put it, the "broad pattern of human palaeontology" had started "to look predictable." After Flores, as his colleague Mike

Morwood admitted, "challenges" arose to "existing notions of what it is to be human the most."[32]

The Hobbit was unexpected—not inconceivable or inexplicable, perhaps, but a sensation nonetheless. It makes probable the discovery of other, similar hominins also isolated in the archipelago, and elsewhere. Still, even as it defined prevailing ideas, the find confirmed a typology of discussion in which those ideas might be revealed, debated, and accepted, a pattern laid down 150 years before. The Flores Hobbit replayed the narrative of the Neander skull both in terms of discovery and in the terms of debate within the scientific establishment: a new species or a degenerate representative of contemporaries; a missing link, or no link at all; an isolated population that lived and died apart from modern contact, or a group that had to interact with sapiens; a disjunction between brain size and cultural behavior; the intrusion of nationalism into disciplinary schools; the hoarding of critical data—the old topics and the manner of their discussion revived. It seemed that, at least with regard to the history of ideas, Haeckel's formulation was proving right. The ontogeny of discourse about Flores was recapitulating the phylogeny of ancestral discovery overall.

Isaac Newton uttered one of the master phrases of the scientific revolution when, with apparent modesty, he proclaimed that if he saw further than other men, it was because he stood on the shoulders of giants. In truth, the phrase had an ancient origin, and within a few years of Newton's *Principia Mathematica,* as scholars debated the relative merits of ancient and modern learning, William Temple restated the epigram to read that moderns were dwarves who stood on the shoulders of "ancient gyants." Three hundred years later paleoanthropology found that its latest advance in learning came from placing a Flores dwarf on the shoulders of Neanderthals.

The early twenty-first-century excavations had targeted Flores because it was a possible route from Asia to Australia for early sapiens. Exploratory efforts turned up stone tools (dated to 840,000 years ago) that suggested previous colonization by erectines; they found floresientines instead; and thereafter nothing seemed to fit except the inherited protocol for discussion. The specimens showed, in Richard Klein's words, a "mix of primitive, derived, and unique features" that did not align with any existing phylogenies. Each trait that fit one model seemed at odds with other traits that fit other models. It was as though a legend had come to life.[33]

From the onset, then, two theses competed. The most striking argued for a new species, that *Homo floresiensis* had evolved from a founding cohort of erectines that had subsequently evolved and speciated and undergone island dwarfism. Islands did that to colonizing populations; small creatures could become large, and large, small. Hippos and elephants had miniaturized rapidly on Crete, Sicily, and Malta. On Flores, rats and lizards swelled up into monster mammals and Komodo dragons, while Stegodonts, an Asian elephantid, and hominins shrank. Along the way floresientines lost some abilities to run, acquired longer arms, grew sapientlike teeth, and shed half the cranial capacity of even the erectines. The loss of brains was not simply a matter of mass relative to body size, since it was proportionately tinier. Yet the Flores brain, like the body it inhabited, seems to have evolved in parts, which left its cognitive capacities intact, as manifest in the ability to make complex tools and tend fire, and perhaps speak. Evolutionary convergence may have taken the form of behavior rather than anatomy.

For advocates, plenty of uncertainties existed. Additional finds—between six to nine individuals—have confirmed the original discovery while deepening the anomalies, among them the mystery of original colonization. After an early episode, there were apparently no further systematic erectine contacts, which would have swamped the gene pool; the next encounter was with sapients, perhaps around 55,000 years ago. Flores may thus have been a kind of Easter Island for erectines, one they could reach but not repeat. A further uncertainty is when the floresientines died out. The community at Liang Bua succumbed to a volcanic eruption around 12,000 years ago.

Other groups may have survived elsewhere on the island, and even into the era of Portuguese and Dutch contact, and possibly to the eve of the Industrial Revolution. It's an astonishing prospect.[34]

The counterarguments began with the assertion that the original specimen represents a degenerate modern human. In particular, the Liang Bua cranium was that of a microcephalic, which would explain a brain smaller than dwarfing alone could account for. Alternatively, it was proposed that the floresientines suffered from Laron syndrome, a genetic disorder. Or, again, others have suggested that the remains testify to endemic cretinism from congenital hypothyroidism, which afflicts some local contemporary populations. Instead of leading to speciation, isolation had enforced a kind of genetic drift that led to pathological dwarfism. The latest hominin find thus uncannily echoed the first: the microcephalic Little Lady of Flores, as journalists called her, was the successor to the arthritic Cossack deserter who crawled into a Neander Valley cave.[35]

The arguments swirl back and forth. For a while, Teuku Jacob, chief paleontologist at Gadjah Mada University, a major power in Indonesian paleoanthropology and a proponent of the microcephalic argument, physically hauled off the specimens and had Liang Bua closed to researchers. The bones were damaged when they were returned, and only with Jacob's death was excavation renewed. Apart from personalities and competing institutions, nationalism had become a presence. It was "very important for Indonesian society," affirmed R. P. Soejono, the lead Indonesian at the site. Flores did for Indonesia, a recent democracy, what Hadar, with its discovery of Lucy, had done for Ethiopia.[36] It created a similar niche and nationalistic legitimization of potential fossil data.

There seems, in brief, to be enough evidence to confirm that the original cranium was not an individual aberration, but there is not enough for the community of scholars to rewrite the received standard version of hominin evolution. More and more the evidence suggests a separate species, but it also accents the degree to which Flores is an outlier, intellectually as well as geographically. A new hominin find in Africa could slide into existing narratives; another hobbit from the Spice Isles would emphasize the range of the possible rather than the realms of the plausible or the necessary. Like the Neander skull, the Flores find did not fit easily anywhere—did not fill gaps of data or narrative. It remained entirely possible for an authority like Richard Klein to sniff that "even if *H. floresiensis* is eventually upheld," it would be as "a curiosity" whose existence would "not alter the basic pattern of human evolution" worked out to date.[37]

Whether the two episodes are versions of a type discourse or an example of historical convergence, there is one colossal difference between them. The Neander skull was found accidentally by quarrymen, then delivered to a professor who tried to explain it without an intellectual system that assumed human evolution. The Flores skull was excavated by a caste of professional archaeologists who were actively searching for evidence of early hominins on Flores. Whether or not progress was apparent in hominin evolution, it was certainly evident within its study. That was the difference between the ancients and the moderns of paleoanthropology. The finders of the dwarves were, in truth, standing on the shoulders of the finders of giants.

Coexistences

For many commentators the most startling upshot was the realization that floresientines and sapiens had occupied Flores at the same time. This did more than upset the received chronology of archaic and modern hominins. It affected the contemplation of floresientines by recent sapiens, because what the record showed might be the result not simply of autonomous evolution but of interaction. Sapiens later occupied Liang Bua cave; they might be a source of floresientine tools; they might be responsible for the extinction of the hominin hobbits. Because an encounter was relatively recent, some contact was unavoidable. Because it was prehistoric, there were no written records, although there might be other artifactual records, among them oral traditions and myths.

Such speculations evoked memories of lurid travelers' tales, wild theories, and epic just-so stories, as well as upset researchers keen to anchor their discipline in the hard sciences. But from the onset, such legends were as embedded in the soils of Liang Bua as lithic flakes and hominin teeth. One of the Australian discoverers, Richard Roberts, reported the stories of "little, hairy people" known to contemporary Florines as *ebu gogo* (the grandmother who eats everything).[38] The legend of *ebu gogo* was well established long before the excavations, and had been investigated by anthropologists interested generally in "wildmen" mythology and by ethnographers attracted to Flores specifically. A reconstruction blending legend, genealogy, and oral and archival history suggests that the arrival of sapiens pushed floresientines to marginal sites, not unlike the case of Pygmies in Africa; that the last hobbit may have survived until 1750–1820, the era of Dutch East India Company's commercial penetration; that direct Dutch administration did not arrive until the early twentieth century. The legends of *ebu gogo*—or the hairy aborigines known variously to other tribes on the island and to islanders throughout the archipelago—could easily have survived to the period of excavation. Oral accounts further describe an uneasy coexistence among island inhabitants until the *ebu gogo* went too far and moderns wiped them out. As *National Geographic* enthusiastically reported in October 2004 and published in 2005, "It's breathtaking to think that modern humans may still have a folk memory of sharing the planet with another species of human, like us but unfathomably different."[39]

What appears to be a broad-spectrum mix of the archaic and the modern may be the defining feature of *Homo floresiensis*. But that same mingling may describe its conceptual history as well, as the material record meets cultural prisms and as science meshes with something like the humanities. In his essay "When Myth Becomes History," Claude Lévi-Strauss once described two ways in which "collections" of things might find meaning. He was particularly concerned with myths, but such collections could be anything. For paleoanthropology, they traditionally consist of data points, lithics, bones, hearths, and strata. They could as easily include story fragments, flakes of memory, and relic archives.

One interpretation is that the collection once had a "coherent order, like a kind of saga," and that the sherds and phrases we find now is "the result of a process of deterioration and disorganization; we can only find scattered elements of what was, earlier, a meaningful whole." All that remain are fragments of a once internal order, and the task is to reconstruct that former wholeness. An alternative interpretation is that the disconnected state is "the archaic one," and that coherence is the product of "wise men and philosophers" who compose order at a later time. The pieces were, in their origin, never part of a whole, which came later from the outside. An example of the first might be anthologies of myths among some North American pre-Columbian peoples; the exemplar of the latter, the Bible. The "gap" between our conceptions of mythology and history, he concluded, might be sealed "by studying histories which are conceived as not at all separated from but as a continuation of mythology."[40]

In a nutshell that synopsis is what appears to have happened at Flores. Historical facts became myth, which then reverted back to fact. While scientists will likely restate this process as simply an expression of a ceaseless dialectic between evidence and hypothesis, humanists will recognize it as a dynamic, also endless, of evidence and narrative. In this sense the Flores story distills the larger enterprise by which we attempt to explain human origins. Whether a master narrative lies in an antiquity of which we have only fragments or whether it lies in a future construction by ourselves, we will likely need to move between myth, science, and history.

The discoveries on Flores, if sustained, have placed ancient hominins across the Wallace Line, a barrier that had been assumed impermeable to hominins until the advent of modern sapiens. But those discoveries have also rafted the discussion across a no less formidable intellectual line between science and other scholarships, one that practitioners had asserted only science could breach: that history, folklore, and metaphysics, like Sumatran tigers and Kalimantan orangutans, however majestic, lacked the wherewithal to master the passage. The floresientines suggested otherwise, which is partly what made them, instantly, a charismatic metafauna. But whether sustained or not, the character of the debate over them has a long ancestry. The dwarves stand indeed on the shoulders of gyants.

Chapter 8

The Ancients and
the Moderns

Each to each a looking-glass
Reflects the other that doth pass.

—Charles Cooley, *Human Nature and the Social Order* (1902)

WITH THE LAST GLACIAL MAXIMUM, roughly 20,000 years ago, the Pleistocene approached a climax that is also a conclusion. Once that crest passed, the epoch underwent no further cycles of ice or hominins; closure is both thematic and aesthetic. The three braiding narratives each acquired a new logic that meant none of them any longer existed outside human influence. Hominins shaped them as much as they shaped hominins. Modern humanity alters climate, tinkers with genomes, and endlessly alters the terms by which it comments on its acts.

How that happened and what it means—the sudden spillage of sapiens over the globe—has sparked a vigorous debate. One group dismisses the planetary sprawl of the sapiens as an accident, the fortuitous collusion of geography and genome. Moderns were no different from earlier hominins, or from beetles and bovids. Another group, however, sees the need for a distinctive, even unique, explanation. After all, bottlenecks had occurred before; populations had previously been squeezed and released; climate had swirled in cycles and undergone defibrillating shocks. But something different seems to have happened with this last outburst that calls for the introduction of some novel feature, even a kind of special creation. What makes the quest yet more compelling is that the evolutionary shift is no longer between species of hominins but variations among one, the sapiens. In some defining way, it is believed, "modern" sapiens differ from "archaic" ones.

What threshold did the modern sapiens cross that allowed for their outpouring? Attention has focused not on their outward appearance, as recorded in bones, but on their behavior, particularly the operation of their minds as recorded in associated artifacts, which change even though the bones of their makers do not. The modern sapiens, it is argued, have a more robust capacity to think, reason abstractly, invent, express, and create. They can substitute their own devices for nature's largesse, and this way rewrite both the geographic and genomic narratives. They can not only make tools to replace talons, but fashion displays that serve like peacocks' tails and emblems that indicate status and hierarchy, or ideas that make possible such inventions and emblems. They have, in brief, culture.

By substituting culture for nature as a guiding principle, researchers apparently finessed the controversy. In reality, they only replaced one variable with another, since culture as an ontological category is even less secure than nature, and with consequences that they surely did not intend, they removed science as an exclusive means of interpreting the difference. In what modernists might see as a burst of irony, natural selection apparently selected for traits that allowed humans to overrule it. Similarly, a self-conscious science, anxious to pare away anything that it could not verify, had to turn to art for explanation. Pleistocene studies found itself climbing an Escher stairway in which each tier of explanation led to another that circled back as it moved up, trudging from nature to culture, which led to nature, which led to culture, and so on. This time, instead of comparing humans to other creatures, the Pleistocene pendulum forced humanity to compare itself with its own image.

Sapients in the Looking Glass

The most recent of hominins, the most closely scrutinized, the most pervasive and powerful—any of these traits would justify *Homo sapiens* as a culmination of things Pleistocene. But of course the real justification for the obsession is that people are studying themselves. For much of Pleistocene studies the sapiens have been the beginning and end—the alpha species in what has become a top-down eco-system and the omega to which the Pleistocene seemed to lead.

It also means that there will be no intellectual wiggle room left—no gaps or looseness—in defining what makes a modern human and what doesn't. The tendency, then, is to stretch out and subdivide that final transition in ways that permit more gradations and yet allow, in the end, for a qualitative difference. Unsurprisingly—inevitably, if Claude Lévi-Strauss is right that the human mind operates by dichotomizing—two schools of thought have emerged. One stresses an abrupt change in behavior as the distinguishing feature between "archaic" and "modern" humans. A mutation in the geographic or genomic settings caused the narrative to jump. The other insists that evolution remained continuous and gradual, that archaic sapiens were fully human, and that an appeal to intangibles like behavior only reinstates the soul that science has struggled to banish from the world machine.

In the realm of formal logic, thesis and antithesis resolve into synthesis. In the realm of real-world scholarship, however, most major problems are rarely solved at all in any technical sense when (as here) the question has a scientific component but not a scientific answer. As philosopher John Dewey observed, the interested community simply moves on to other business. Such transitions typically follow a generational pattern. One generation quits trying to square its predecessor's circle. Whether or not the concept of a human revolution can apply to Middle Stone Age hominins, it certainly characterizes a phase of contemporary inquiry about them.

The Saga of the Sapiens

As the Pleistocene entered its latest and likely among its most savage glaciations, the last glacial maximum acted like a broom and a vise, squeezing Europe in particular between ice and mountains. This final cycling seemed to miniaturize the long history of the epoch and the saga of *Homo*. When it ended, the Pleistocene concluded its framing of human history as it had begun it, with a single species.

The earliest sapiens had emerged, arguably around 200,000 years ago, at a time when the Earth had a full complement of hominins. There were several varieties of sapiens in Africa, while outside it the heidelbergentines and Neanderthals claimed Europe and the Middle East, and the erectines still held their ground throughout Asia, along with whatever species had diverged from them. Evolutionarily, there was nothing extraordinary about adding another member to the clade. What distinguished this era from the rest is that Earth would apparently add no others. The sapiens removed all rivals, not only among hominins but among mammals generally; only those megafauna it tamed or tolerated survived.

The earliest tendencies toward the modern sapiens probably appeared 500,000 years ago in Africa with clear definitions both anatomical and genetic evident between 100,000 and 200,000 years ago. Like their ancestors they soon took to their heels. During the interglacial prior to the last glacial maximum, these early sapiens expanded into the Levant before being pushed out when the glacial climate returned. They replaced Neanderthals during the warm phase; the Neanderthals returned during the cold. The different outcomes seem to lie in somatic distinctions in that the Neanderthals were better adapted to cold, while the sapiens remained a more tropical species. Otherwise they occupied similar niches with similar technologies. "Initially," as Richard Klein observes, "the behavioral capabilities of early modern or near-modern Africans differed little from those of the Neanderthals." Each species came and went as the larger circumstances allowed.[1]

Yet the sapiens' success by the latter part of the Pleistocene was a close-run thing. More than once bottlenecks had threatened to choke off the sprawling hominins. At such times the geographic narrative colluded with the genomic to shut down paths outward, to squeeze refugia, and to drain the gene pool. Such episodes went beyond the usual suspects acting in the usual way to remove bridges and put up barriers to migration—beyond intensified droughts, sharpened cold snaps, quickened swirls of sand and dust, all of which had happened before and would happen again and again. Throughout the Pleistocene, while Milankovitch rhythms had cycled climate in predictable ways, the ranks of the sapients had continued to swell and shrink with the scope of their larger setting. Early sapients had pushed across Sinai into the Levant, and then retreated when glacial conditions returned. The sapients seemed little more than yet another by-product of an endless churning of hominins, and were vulnerable to exceptional events that might either drive them to extinction or liberate them. Something extraordinary would have to lift the standard story line out of its usual ruts.

Two somethings apparently did, as current research reads the evidence. Both were, in their own ways, supereruptions in the sense that they were huge, sudden, and anomalous. One narrowed options and led to a sapient retreat, and the other flung the portals open to allow a sapient sprawl.

The squeeze took the form of a monstrous eruption of Mount Toba on Sumatra between 73,000 and 71,000 years ago that smothered much of southeast Asia and stunned the global climate. The Toba eruption was a monster—an order of magnitude greater than any other known in the Pleistocene and the second largest recorded for the Phanerozoic era. For six years it is calculated to have plunged the Earth's climate into winter; for the thousand years that followed there was a deeper cold spell than the previous glacial maximum. It may have ecologically damaged much of southern Asia, the prime habitat of the erectines. Its fallout in terms of lost habitat and likely famine may have pushed *Homo sapiens* to the brink.[2]

Unlike Neanderthals, sapients needed a tropical refugium, and Africa, their hearth, offered the largest available. The genomic record suggests that the population shrank, and it may have collapsed. Estimates of survivor numbers run from as low as forty individuals for two centuries to extant tribes that collectively held ten thousand reproductive females. The event—or some combination of crises that it helped catalyze—cleft a deep scar in the hominin genome. It bequeathed a freshened genetic legacy for the sapiens, a recency of renewal along some six principal lineages; it left Africans with a greater genetic diversity than any other cluster of sapiens, partly from the population size of survivors and partly from longevity; it sculpted a species that, when circumstances rebounded, found an ideal arena for dispersion. Whatever effect the Mount Toba super-eruption actually had on the geography and genome of the sapiens, a surpereruption of sapiens followed.[3]

A rapid radiation of sapiens from Africa ensued, like the creatures loosed from Pandora's box. It first stirred around 60,000 to 70,000 years ago with an amelioration of geographic conditions before spilling out of Africa, and it quickened around 40,000 to 45,000 years ago as some new feature seemed to amplify human firepower. By 55,000 years ago sapiens had overrun southern Asia and crossed into Sahul and Australia. By 35,000 they had punched into Europe, tracking the receding ice and crowding the Neanderthals. By 20,000, the height of the glacial maximum, they had crossed Beringia into North America. By 10,000 they had reached the southern tip of South America. They had spread with the implacability of a plague or the Norway rat, filling up one new island or continent after another. Even as the Pleistocene melted into the Holocene, they were still probing and pushing.

This much most researchers agree on. The interpretation of what happened next has split into two general camps, a dialectic that has not so much found a synthesis as left most practitioners in a muddled middle. One camp argues for a human revolution that produced, even among the sapiens, a variant whose behavior was distinctly different from those who appeared earlier. Although they display no anatomical distinctions, modern humans acted differently from archaic humans; they possessed a radicalized culture in the usual sense. The other camp dismisses the plea for a revolution as romantic nonsense. There was no distinctive, quantumlike mutation, only a continuation of incremental gradations. The argument for a revolution, they insist, is a misreading of the evidence and probably a by-product of a cultural "privileging" of Eurocentric norms.[4]

Yet that kind of bias is inevitable, and it appears in both camps. The controversy forces people living in an analogue world to make a digital (binary) choice; shifting from natural criteria to cultural ones only defers or disguises that decision. Whatever the distinguishing "cultural" differences between early and later sapiens, the real distinction lies within the minds of those observing them.

Thesis: The Art of Human Revolution

In the first scenario, proponents of a human revolution argue that by the time sapiens returned to the Levant, around 60,000 years ago, they were distinct from—behaved differently than—those predecessors who had been driven out by the glacial cycle. The issue hinges on whether, in fact, such a change occurred, and if it did, what could have caused it.

There seems to be no outside stimulus adequate to jolt the sapiens into a radically new mode of life. Other than the occasional shock of a Toba-like volcanic winter there was no fundamental change in the cadences of climate. The next glaciation followed according to the usual scenario, since the incident at Toba did not alter the deeper cosmological cycling. Nor did there appear to be any anatomical changes in *Homo sapiens*. Compared to contemporaries, the skulls seem more archaic but not sufficiently distinct to warrant identification as a new species. The population—the gene pool—was squeezed and sieved, not altered in its basics. There was no morphological alteration adequate to account for successful behavioral change. That meant commentators had to look outside the parameters of the geographic and genomic narratives that had so far served to define the Pleistocene. They turned instead to behavior itself or, more specifically, to cognition. "Perhaps because of a neurological change" early sapiens "developed a capacity for culture that gave them a clear adaptive advantage over the Neanderthals and all other nonmodern people." The revolutionary, newly wired sapiens burst into the Levant and then beyond.[5]

The solution to the question of what happened is to distinguish between anatomically and behaviorally modern humans. The sapiens differed from earlier versions because they had something that did not change their bones. The archaics had been "cognitively human," as Klein put it, "but not cognitively modern." Only when they acquired "the fully modern capacity for culture" did they obtain "an adaptive advantage over their archaic Eurasian contemporaries." What did such a capacity bring? It allowed refinements in the technologies of stone, bone, and wood; better language, communication skills, and social organization; an appreciation of beauty; a talent for manipulating symbols; an interest in ornamentation; an imagination capable of devising and testing ideas as evolution did mutations. An anatomically modern human could make spearheads to hunt; a behaviorally modern human could paint hunting scenes on the walls of caves. To simplify a lot of waffling jargon, the change sparked a mind that could produce art and, later, science.

The paradox is palpable. To explain modern humanity the sciences of human origins turned to art for answers. They found distinguishing artifacts that were ornaments not designed to hack, scrape, or puncture. They defied instrumental interpretations, for they were obviously art or objects endowed with symbolic significance. The "motifs," such as those found at Blombos Cave in coastal South Africa, suggested "arbitrary conventions unrelated to reality-based cognition." So while the objects spoke a language of symbolism, they were themselves not to be spoken about through the lenses of art history and the philosophy of aesthetics.[6]

"Culture" had ambiguities of its own, however, and it was particularly treacherous when used to array peoples into hierarchies of the sort favored by paleoanthropologists. After all, the thrust of twentieth-century anthropology, as organized by the Boasian school, had pleaded powerfully against ranking cultures along an axis of primitive to civilized states, arguing that other than their technological richness, all cultures possessed the same cultural capacity, and that they were equally human, whether they used machetes or chain saws or, in Pleistocene context, threw spears by hand or used an atlatl.

Most commentators identified two possibilities. One held that sapiens had possessed the potential for such cultural expressions but awaited some catalyst, "some momentous social or demographic change," before they were able to express it. Like Dorothy with her ruby slippers, they had the power all along and didn't know it. The other possibility is that some somatic mutation in the neural network allowed for a change in behavior. Sapiens acquired a more robust cortex in the same way Neanderthals built up muscle, and by a similar mechanism: organic evolution, in this case with the neural network rewired by natural selection instead of just bulking up its mass. The first explanation emphasized the mind, the second, the brain. The consensus assumption was that the latter made possible the former.[7]

Both left a gap in the record: a new missing link, this time of the dark matter of culture. Researchers sought to fill it in with an evolutionary sequence of artistic expression—the first "human" speech; the initial use of decorative ochre; burials with flowers; a necklace; a flute; carvings; drawings; paintings; figurines—the list goes on and on. All seem to burst on the scene around 35,000 years ago. But what about the earlier sapiens? When did separation occur? Over how many millennia? And what about those who swarmed out of Africa around 50,000 to 60,000 years ago but left no equivalent galleries of art? If behavior is what distinguishes archaic from modern humans, and certain cultural expressions are an index for behavior, then were those earlier emissaries truly modern or only intermediaries? If not fully modern, then what about their descendants? If modern, then how do they differ from those hominins who are only anatomically modern?

In short, the behavior argument proposed another suite of questions to replace those it had vanquished from the anatomy argument, and it left researchers struggling to keep it within the disciplinary bounds of science by calling it "symbolic behavior" or "behavioral" modernity instead of "painting," "thinking," or "inventing." In fact there was less likelihood that art would become science than that scientists, struggling to impose order on scanty and refractory data, would turn their science into art. They could create context and meaning by how they displayed their artifacts, with their aesthetic sense, not unlike those early sapients who arranged their dead for burial and decorated the scene with made objects. Whether or not the narrative might have a satisfactory thematic closure, it could have a pleasingly aesthetic one.

Antithesis: The Counterculture of Continuity

An antithesis to the revolutionary one emerged. Instead of splitting the sapients, this counter-thesis urges a lumping. It dismisses the apparent quickening of human behavior around 40,000 to 50,000 years ago as an accident of the fossil record and a bias by its advocates. Look more closely, critics argue, and you will find a graduated continuity of artifacts and behaviors and a suite of sapients scattered across space and time. Instead, the counter-thesis proposes a "gradual assembling of the package of modern human behaviors in Africa, and its later export to other regions of the Old World." In place of a neural revolution the critics are inclined to identify missing links—complementary species to the heidelbergentines, among them a proposed *Homo helmei*—that can maintain the old order.[8]

In any debate it is almost always easier to criticize the proposal of others than to defend one's own. Inserting new links in order to maintain the integrity of the hominin chain only shifts the location of unanswered questions, and of course it brings its own cultural baggage. Halving the difference between existing morphologies does not get them from one state to another any more than halving the sides of a square eventually gets to a circle. In Zeno's famous paradox, Achilles is so swift that at any instant he can cover half the distance between himself and the tortoise ahead of him, yet he will never catch up; by similar reasoning, *Homo heidelbergensis* et al., however swift, can never catch up with the sapients. Mathematics evades the issue by invoking the concept of limit. The equivalent for hominin evolution would seem to be an unlimited culture.

In proposing continuity the contras have deeper objections than that their rivals fail to explain the artifactual record. They believe that the revolutionary manifesto harbors several larger toxic corollaries. It creates—perhaps unwittingly—a "gulf separating humans from the rest of the biological world." No such appeal to inventive or artistic culture is necessary to work through the phylogeny of the chimp or the elephantids; none should be necessary for that biological entity known as the hominins. An appeal to intangibles such as art only severs the subject from the natural world, which removes it from the "realm of normal scientific inquiry." Equally, the whole concept of "revolution" is suspect, tainted as a surrogate for the "search for the soul," as a cat's paw for a "profound Eurocentric bias" that wants to find modernity within a European cradle, and as a way to inject European norms into definitions of modern. The controversy, that is, derives from "semantic" issues that hinge on terms like "revolution" and "modern" that date back to the European Renaissance. Such concepts, the contras conclude, are inappropriate when used to describe the "behavior of people prior to the sixteenth century, let alone those of the Pleistocene."[9]

In this last claim, the counterculturalists are surely right because the quarrel does indeed pivot on social and historical considerations. The controversy revives the terms of Renaissance discourse about nature and culture (and on a deeper plane, body and mind) and about the two scholarly cultures that arose to discuss them. But cultural construction applies to the contras no less than to the revolutionaries. Archaeology, paleoanthropology, geomorphology, the geologic timescale, modern science, and scientism—all are creations of a particular society. It is not simply ideas and norms of modern behavior that are Renaissance products. So, too, is the modern science that would look to banish antiquated notions about souls and revolutions and instate a new logic by which to argue. The same controversy after all applies to science. Did it end the Renaissance with a revolution or was its emergence the outcome of incremental improvements in thinking that are continuous from Anaximander through Roger Bacon to Isaac Newton? The entire enterprise by which contemporary societies inquire into human origins and define modernity is a Western invention.

In fascinating ways the dispute over archaic and modern humans echoes that late Renaissance Quarrel between the Ancients and Moderns. The question of what, if anything, differentiated the assorted sapients of the late Pleistocene could be applied equally to those who studied them in the late twentieth century. Was the history of that research itself a testimony to cultural revolutions or to a Burkean conservatism? What differentiated the 'nominally archaic research from the self-consciously modern one? Does positivist science today serve to segregate inquirers as art reputedly segregated sapients in the past?

Battle of the Books, Battle of the Bones

It was the culture war of its time. In 1687, Charles Perrault published "The Century of Louis the Great," in which he celebrated the literature written during the reign of the Sun King and observed that "learned Antiquity" could not speak to contemporary conditions. Bernard le Bovier de Fontenelle immediately agreed, and added his own *Digression sur les anciens et les modernes* in 1688. Those who favored the Ancients—the classicists—promptly rebutted the charges, insisting that the Ancients had identified the grand themes of human life and the genres by which to address them, and that culture could best advance by their continued study and emulation. Thus began what became known as the Quarrel of the Ancients and the Moderns.

The dispute spilled over the English Channel in 1690 when Sir William Temple published "An Essay Upon the Ancient and Modern Learning." Those who converse among "the Old Books," Temple asserted, "will be something hard to please among the New," although they "have their Part too in the leisure of an idle man." Mostly, "they do but trace over the Paths that have been beaten by the Ancients, or Comment, Critick, and Flourish upon them, and are at best but Copies after those Originals." He made an exception for topics that did not exist in ancient times and conceded that modern works "of Story, or Relations of Matter of Fact, have a value from their Substance as much as from their Form." Still, although we know more than the ancients did, "Science and Arts have run their circles, and had their periods in the several Parts of the World." The eternal wisdom endures, and it is to be found in the Ancients. While the Moderns have a place, it is less exalted than their wit proclaims and their jejune criticism merits.[10]

In France the quarrel dragged on. At stake was more than literature, because books were the primary means of conveying knowledge, and it mattered whether new wine could be poured into old bottles or if a new bottle was needed. In Britain the dispute passed like a storm, ending in Jonathan Swift's satirical *A Tale of a Tub* (1704). To it he affixed as a prolegomenon, *The Battle of the Books*, in which he imagines an epic battle between "the Ancient and the Modern Books in St. James's Library." Aesop opens the contest by comparing the Ancients to a productive bee and the Moderns to a consuming spider. Then the books battled, as Homer led the Ancients' horse and Euclid commanded its engineers, and Momus (mockery) and Criticism directed the Moderns. In the end it is Swiftian satire that rules the field.

Such controversies among the learned reoccur, in Swift's observation, "under several names; as disputes, arguments, rejoinders, brief considerations, answers, replies, remarks, reflections, objections, confutations." What made this quarrel different was its historical timing. Perrault wrote his panegyric the same year Isaac Newton published the *Principia Mathematica,* and Swift concluded the controversy with a satire published the same year as Newton's *Opticks.* These were forms of learning that seemed to transcend anything the Ancients had to offer.[11]

Between them those two Newtonian tomes consolidated the transition from Renaissance humanism to modern science and became the exemplars for the new learning. The one described how experimentation could create new knowledge, and the other depicted how to organize new and old data into a coherent, mathematical accounting of the world. Together they proclaimed that something novel had come into the realm of knowledge, a new way for scholars to inquire, that yielded results not known to the ancients nor even possible for them to have acquired. The way to knowledge, call it a change in behavior, had fundamentally shifted in ways that distinguish an archaic culture from a modern one.

In brief, the world was not doomed to endlessly cycle and recycle: it could advance. The emergence of science was the very paradigm for progress, since it showed genuine growth in knowledge and the means to acquire more. It was the signature scholarship for modernity. Immediately it began to shoulder the other, older forms of learning aside. The bottomless book of nature replaced the limited-edition books of the ancients. A text-based scholarship grew arithmetically, as it were, while science grew exponentially. More and more modernity became the expression of the further penetration and hegemony of science and a science-based technology, and that succession is what most completely distinguishes the Ancients from the Moderns.

The internecine quarrel over whether that boost in learning makes a revolution or not still preoccupies scholars. Debates over the validity of a "scientific revolution" thus echo debates over the "human revolution" of the late Pleistocene. But there occurred, over a relatively short period of history, a change not only in degree but in kind. The Ancients read texts; the Moderns read nature with science.[12]

Positivism Comes to the Pleistocene

Pleistocene studies, too, had its quarrels between ancients and moderns, and these harked back to the lineage Swift had roll-called. Since ancient times this inquiry had belonged with species of scholarship that asked who we are and how we should behave, which is to say, with ethics or moral philosophy. It concerned itself with the intangibles of human motivation; it sought answers in the accumulated lore of human experience as recorded in religion, art, literature, philosophy, and accrued knowledge generally. It asked the whys of existence. It wanted, ultimately, to understand the purpose behind it all.

It was a knowledge recorded in books, and the Renaissance—literally, a rebirth—thrived on the recovered texts of the ancients and devised a scholarship to analyze such works. The scientific revolution relocated that focus; the most vigorous inquiries turned to material nature and to immediate rather than final causes. It shifted from pinning down the purpose of human existence to the mechanisms by which humans behaved and, after Darwin, to those mechanisms by which people had descended from predecessor species. Progress came by replacing the antiquarian methods and lore with scientific ones. Bit by bit, scientific information, or what was asserted to be scientific, supplanted inherited explanations and their mode of expression. For a long while, narrative endured, with progress replacing providence as an informing principle. Much as science would reduce the operations of the mind to those of the brain, so the scientific study of human evolution would reduce behavior to biology. Matters of morals and culture were decorative and derivative.

In the post–World War II era this trend quickened, stimulated by government funding, dismay with the old narratives, and the intellectual migration of brainpower from Europe to America that transplanted an analytical philosophy in place of a seemingly sloppy pragmatism. American science, especially, veered sharply in how it sought to explain matters of human origin and evolution. It attempted to replace empirical information with experimental data where possible; it dismissed narrative as a means of analysis and synthesis and turned to descriptions of process alone; and it endorsed a program of implied philosophy, what can only be called a variant of positivism, that was heavily influenced by contemporary research in the philosophy of science that leaned heavily toward logical positivism. The first two shifts characterize Pleistocene studies overall. The study of geological processes, for example, challenged historical geology as the end point of explanation and eventually redefined historical geology away from narrative into chronology. The study of landscapes, to continue, dropped Davisian cycles of erosion, in which every landscape moved inexorably through a life cycle from young to mature to old, and instead studied the dynamics of erosion. The description of the particular erosive process served as an end explanation.

All three trends, however, applied to the disciplines concerned with early hominins. The first reform moved artifacts from the realm of trophies and curiosities, devoid of context, and established their provenance while making them objects suitable for laboratory analysis. Taphonomy and zooarchaeology, for example, could analyze bones in ways shorn of untestable speculation. The second transformed stories into hypotheses. Process could be studied apart from its origins and futures; it just is. Like Newton's laws of motion, it has no need for a narrative frame. The third reform confirmed the discipline as a hard science well and truly divorced from the soft scholarship of the humanities. The study of the mind had no pertinence to modern understanding of behavior based on the brain, or, to put it in terms of battling books, a reading of Homer, Livy, and Plutarch had no more relevance for understanding human behavior than Galen or Hippocrates had for modern medicine.

In this latest reincarnation of the quarrel the moderns again allied with science and left the moldy books of the past in the library. Culture was a suspect term, certainly not a testable argument. If a change of behavior was at the heart of the human revolution, it lay with processes of behavior, not with filmy notions of culture. The future lay with the study of the brain, not the mind. Those ancient writings—however inviting—were so much graffiti on the walls of history.

Mind over Brain

But was, in fact, the mind irrelevant to scientific study? In reviewing the impact of Darwinian thought, Karl Popper concluded that consciousness could have value with respect to natural selection, because it allows for imagined futures in which it is possible to test ideas and behaviors without having to suffer real-world consequences for them. People can replace chance mutation with experiment and natural selection with choice. Evolution, in the sense of change through time, still depended on variety and novelty; but both the sources of change and the selective pruning among them now resided with the mind rather than with chromosomes: "The selection of a kind of behaviour out of a randomly offered repertoire may be an act of choice, even an act of free will." One may choose among random events without the choice being random. Whether or not organic evolution had a purpose, people's lives could. Creativity made possible art. It made possible science. The mind could impose its will on the brain.[13]

This led, in Popper's view, to the enduring question of how mind and brain interact, which returned to T. H. Huxley and A. R. Wallace. Huxley upheld the primacy of the brain. Humans were automata, whose minds could express the actions of, but not themselves act upon, the brain. Thoughts no more influenced neural workings than a "steam whistle" did a locomotive. Mental states are indistinguishable from physical states: psychology and anthropology in the end reduce to biology. By contrast, Wallace had argued for social sensibilities and moral senses as powerful forces that could block, and even replace, the action of natural selection, and although he doesn't cite him directly, Popper upholds this position. He thought the "so-called identity theory of body and mind" in fact to be "incompatible" with the theory of natural selection. He dismissed Huxley's position as "not . . . very interesting" and turned the criticisms once made against Darwin's concept against Huxley's, that it would be best if it were not true, but if it were true, it would be best not to advertise that fact. "All that ever seems to have been said" against the theory that brain and mind interact, he concluded, is "that the universe would be a simpler place by far if we did not have experiences—or since we do have them, if only we could keep mum about them." The mind-body problem was another reincarnation of the culture-nature problem.[14]

Equating behavioral change with neural rewiring, by reducing the mind to the brain, helps keep the topic within the nominal bounds of natural science. But the core issue is a shape-shifter, no more likely to be pinned down than the Norse god Loki. Ever greater refinements and intellectual jugglings are necessary to parse the moderns—be they modern sapiens or modern scholars—from all the rest. Criteria have to exclude Neanderthals and floresientines, and even archaic or anatomically modern humans, without flensing away genuine humans who do not think or act as we do, which is tricky since the "we" tends to be the particular culture of the researchers. They are Westerners by education and intellectuals by avocation. The conversation about modernity takes place within their self-selected group—among people who distinguish themselves from the general population and themselves from one another—by their "cultural attainments." There is a sense in which those evaluating archaic from modern sapiens are projecting their own self-identity onto the past. That may partly explain why *Homo floresiensis* came as such a shock because here was a "non-encephalized descendant" who evidently "arose from an encephalized ancestor." Those doing the research are almost all professors, or at least the product of extensive education, and learning is supposed to go the other way.[15]

It is often noted that, with modern humans, life became conscious. In truth, consciousness is present among many species, but with it humanity can express itself in special ways, among them the ability to imagine other experiences and worlds. This is the basis not only for art, which has become an archaeological token for modernity. It is also the basis for scholarship, including science. In contemplating hominin evolution, the mind is acting on the brain. To deny this might, as Popper suggests, make the world a simpler place, but it means that science may be left in the awkward position of denying itself.

Ochre and Aurochs

Art as abstraction, art as personal expression, art as an index of cultural development—however conceived, art is increasingly accepted as a manifestation of certain kinds of cognition—which distinguishes among hominins. Sapiens could create art, erectines could not. It is also generally assumed that "symbolic societies intentionally communicate cultural identities," which is to say, social groups that can produce art rely on symbols to define who they are. This has proved true also for archaeologists who appeal to preserved symbols to define the groups under their study, for unlike other symbolic practices, art leaves an artifactual record. It can be studied by moderns.[16]

The wall paintings in the great caves at Altamira, Lascaux, Cussac, Chauvet, Niaux, Les Trois-Frères—the Delphis of European prehistory—have done what Plato claimed great art always does. They represent reality and they distort it. They are both mirror and lamp, they reflect and illuminate, but each necessarily represents reality imperfectly as idea or inspiration moves to material expression. Certainly this has been true for the cave art, which has both dazzled observers and deflected the discourse about hominin history in ways that have split the field. Is art a true representation of modernity or, as Plato feared, a distortion that misleads?

There is no way to decipher what the occupants of Altamira thought about government, the supernatural, truth, freedom, and right and wrong or whether they could even speak or otherwise communicate about such abstract concepts; but they left a record in their art about how they saw aurochs, deer, and horses. However complete or incomplete a picture they give of the last glaciation in Europe, they have come to represent the societies that drew them. Whatever the purpose of such art in its times, it has found a sure role in the modern study of those peoples. The controversy over a putative human revolution hinges on the sudden appearance of those paintings in the Pleistocene.

Arts and Sciences

Continued research has closed the gap between simple tools and abstract art or, since "art" is not an operational term in Pleistocene science, "symbolic representation." Evidence of ornamentation and artifacts that have representational rather than utilitarian uses have emerged from sites occupied by early sapiens and far removed from Europe. Blombos Cave, for example, has yielded ochre incised with deliberate designs, shell beads apparently intended as necklaces, and an engraved bone, all dating to 77,000 years ago. There is also evidence of fishing tools that suggest a more complex culture, one that could join simple tools together in ways to perform socially complicated tasks. Or, as one group struggled to characterize the discoveries at Blombos, "[d]epictional or abstract representations, and personal ornaments, are generally considered archeological expressions of modern cognitive abilities and evidence for the acquisition of articulate oral language."[17]

Further discoveries keep pushing the origins of art, like the origins of modern humans, back in time. The art is simple: drilled shells, beads formed into necklaces, engraved ochre, carved figurines, patterns etched on bones or stone that do not represent anything natural but exist as emblems. Surely too there is an aesthetic sense at work in the progressively fine flaking of stone, and in the shaping of bone into fishhooks, all of which merge beauty with utility. As other aspects of the culture matured, so did art, becoming more elaborate until in glaciated Europe around 40,000 years ago an extraordinary gallery of cave paintings suddenly effloresced in what may be the first outbreak of a European artistic tradition. Cave paintings filled France and the Pyrenees much as cathedrals did medieval Christendom or as Cubist paintings stocked museums in the twentieth century.

The earliest discovery happened at Altamira Cave in 1879, but as with fossils, so with fossil art—confirmation emerged slowly and grudgingly. It did not seem possible that such spectacular drawings could come from peoples older than even known primitive tribes. Over the years more caves revealed similar paintings—150 sites in all—and the dates kept receding deeper into the Pleistocene. Some of the drawings are as old as 35,000 years, although most are more recent, reaching a climax after the last glacial maximum. Refinements of dates and techniques have come into a consensual equilibrium that places them in the period known as the Magdalenian, 10,000 to 18,000 years ago. The rock galleries contain handprints from blown paint, finger etchings, some engravings; but mostly the walls radiate drawings of the era's megafauna. Aurochs, red deer, ibex, and horses dominate, but there are cattle, birds, felines, bears, rhinos, mammoths, lions, and human hunters as well, and some smatterings that appear as abstract emblems. The paints are a mineral pigment, usually mixed with charcoal, and sometimes with plant oils to extend and bind. It was applied variously by brush or blowing. Some overlie in a Paleolithic palimpsest.

The images are broadly naturalistic. They are sufficiently accurate that there is little doubt what species is drawn, yet there are enough liberties taken to express a sense of personality and style and added value. An exaggerated and extended horn and oversized antlers transform what nature offers into what the artist finds significant. The species reproduced are sufficiently distinct that they can help date the various compositions by offering a correlation with relic bones. Interestingly, they stand alone, as representations of the animal or hand by itself, without any landscape setting.

But one can also ask, How accurate is the art as a representation of the society that produced it? Here similar issues of exaggeration apply. Societies express themselves variously, of which cave paintings are only one possible means. They may invest their imagination into producing better microflakes or organizing group hunts or through storytelling. As expressions of their sustaining society, the paintings may resemble the hulking hump of an auroch, a way of conveying a telling trait but not necessarily a defining one. The paintings lack any social setting. Their habitat is the cave, and they are usually in a deep recess of the cavern complex not frequented in quotidian life.

Art has its own logic and styles of appreciative scholarship. While aesthetic fashion can group into periods, it is not chronologically diagnostic in the way radiometric dating is. Art is not sequence-dependent, as is flint-knapping technology. Nor is it judged by how accurately it replicates the material appearance of the original. Moreover, the representation of an auroch must subsequently be itself represented through the interpretation of the observer—the eye of the beholder. What is archaic in art and what is modern lacks a fixed referent. Interestingly, the authenticity of the Altamira paintings was accepted at roughly the same time that Pablo Picasso, a later native from the Pyrenees, was breaking down the orthodoxy of representational art and kick-starting modernism, in part by reviving nominally primitive painting. The modern art of the past century more resembles the bulls and bison of Altamira than it does the grand manner of the eighteenth century or the impressionism of the nineteenth.

Such artifacts are nonsense if viewed as data in the same sense as bones or lithics. They are not tools after the fashion of a bone awl or a fishhook, but exist to elicit awe, feeling, idea, a response from the viewer. Science can analyze the source and chemistry of the pigments on the wall as it can analyze the cause of scratch marks on a femur; it cannot deduce the source of inspiration or the effect on the observer. Art as an index of a broad culture is not a scientific measure, and appealing to it can only unbalance a scientific explanation. But something similar happens by putting people into a definition of the Pleistocene, because people are only partially explicable by the methods of science that can analyze DNA, the depositional history of bones, and the geographic radiation of fishhooks and incised ochre, but cannot invest those objects with the kind of significance that its sustaining society wishes. For that understanding researchers must also turn to history, philosophy, religion, and art, even as art makes a fickle index of culture and culture makes a troublesome register of what makes sapients modern.

Self and Other

There is another mirror to reality, the one we see in those around us, what Charles Cooley called "the looking-glass self" and what George Herbert Mead identified as the significant others against whom we come to know ourselves. The self becomes an object first to others and then to the self observing those others, converting I into me through the absorption of a "generalized other." The self comes to awareness through a social mirror by seeing itself as it understands others see it. The presence of evolutionary ancestors can serve as just such a social group.

People not only see the past through themselves, they see themselves through the past. In reflecting upon the Pleistocene and its hominins, they get reflected back. The habilines, the erectines, the heidelbergentines, the Neanderthals, the archaic and not-quite-archaic and almost-modern sapiens all form a historic community by which to-day's humanity comes to define itself. Those others form a collectivity of foils, or dichotomies, or gallery of looking glasses out of which the self can create not only an image but a story. Like Magdalenian art, the outcome has distortions, and like much modern literature, an unreliable, or at least an unstable, narrator as each reflects on the other. As Jonathan Swift understood, such self-absorption can lead to solipsism or satire, which he defined as "a sort of glass, wherein beholders do generally discover everybody's face but their own."[18]

The way to avoid that self-reflection is to have a mirror outside the control of the person looking into it. For Pleistocene studies such a project requires a geography and a genome beyond meddling by the observing sapiens or else the looking-glass self may end in a hall of mutually facing mirrors, each reflecting the other, and the other, and the other, until the story ultimately dissolves into nothing more than the act of storytelling, since even the narrative has no fixed end. Yet this is what modern humanity has done. It has so tinkered with its surroundings that even the climate is changing, and so meddled with its own genome that there is no longer an objective Other against which to measure.

Many of the most powerful minds that have pondered the evolution of humanity have spoken of Earthly life, through people, becoming conscious of itself. For many such scholars an expression of consciousness is the most transcendent of nature's achievements, for it allows the outcome to ponder its own existence. The sentiment emanates not only from quasi-mystics like Teilhard de Chardin, but from evolutionary theorists like Julian Huxley and Theodosius Dobzhansky, and from paleoanthropologists like Phillip Tobias. Even avowed secularists seem to accept a kind of transubstantiation by which matter becomes mind.[19]

But what is being reflected? And through what representations? And by what means might it be understood and discussed? The difficulty with an appeal to cave painting as a defining feature of modern humanity is not with the art but with the science, or rather with the attempt to absorb art within science, to insist that science and science alone is adequate to the task of explaining what makes us what we are. In his interrogation on natural selection, Karl Popper argued that "we have to admit that the universe is creative, or inventive. At any rate, it is creative in the sense in which great poets, great artists, and great scientists are creative." He concluded that "it is important to realize that science does not make assertions about ultimate questions—about the riddles of existence, or about man's task in this world." Or to put the matter in terms of the Pleistocene, a history that excludes the ochre and aurochs would be a crisper chronicle but an inaccurate one. To see why, consider the contemporary histories of those caves that have become so vital to the definition of modern humans.[20]

Lascaux II

Lascaux Cave was discovered in 1940 by accident when a group of children and a dog found a hole made by a fallen tree in the Vézère Valley. As they probed the opening, they found that it led to a cavern, and then through rooms and corridors—a main hall and galleries, then a seemingly cathedral-like plan with apse and nave, the walls shimmering with paintings and a few engravings of a Pleistocene menagerie, nearly two thousand figures in all, as seemingly fresh as the day they were created. One critic observed that the images "bridged the distance between then and now" and seemed to dissolve time itself. In the years following World War II, the full splendor of the scene became known, mesmerizing archaeologists, art historians, and the public equally. The Cave of Lascaux became the world's most celebrated museum of Magdalenian art and as much an icon of the Pleistocene as mammoths and Neanderthals.[21]

The discovery revolutionized thinking across disciplines. Art, it seems, did not begin with the ancients of Europe, with the sudden apparition of a Greek miracle. It had coevolved with *Homo*, and their mutual origins kept being pushed further into the past. Everything was older than previously thought. Creativity did not evolve according to a developmental template by which hand axes morphed into spears and then into rifles. From the perspective of art critics, "nothing supports the contention that we are greater than they." On exiting the cave, Pablo Picasso declared, "We have invented nothing." But once established as legitimate, the decorated caves, of which Lascaux is but one of a score in the Vézère, began to warp the intellectual geography of Pleistocene studies.[22]

The Real Becomes the Ideal

The paintings at Lascaux have suffered an odd fate. Their revelation brought attention, and with attention popular appeal and visitation. Then, what the ancients had created, the harsh lights (and heavy breathing) of moderns put at risk.

Public access rose steeply after the war. By the mid-1950s, over twelve hundred visitors a day were introducing the vapors, lamps, and enthusiasms of modernity into the chambers of the Lascaux labyrinth. What ten millennia of semisealed geology had preserved, a decade of trampling, observing, photographing, and breathing began to decay. The paintings visibly deteriorated, so much so that entry was closed to the public in 1963, the paintings were restored by an art historian, and they have since been monitored by curators. The public and scholars both knew the art through representations of it, primarily by careful photographs; but that was nothing unusual, for much of early hominin history had been known through casts and reconstructions. Even the most august authorities, like Huxley, pronounced on Neanderthals on the basis of plaster casts. They dealt with simulacra.

In 1983 this process—making art from art—intensified when a replica of the two major chambers, the Great Hall of the Bulls and the Painted Gallery, was built some 220 yards away. Lascaux II hovered in a cultural Lagrangian point somewhere between a museum's collection of study casts, an Epcot Center, a Biosphere 2, and virtual reality. What the re-creations added was not simply the images but their simulated setting. In order to replicate the original, the Lascaux artists had even incorporated the irregularities of the stone ceilings and walls into the pictures, such that the hump of a bison might overlie a lump of rock. The re-creators made a cast of a cast. Lascaux II did to Lascaux what the images on Lascaux's walls did to the creatures of the late Pleistocene. The representation was itself represented.[23]

Yet there was little option. Previous attempts to preserve major artwork, such as reconstructing the Parthenon and repeatedly cleaning the *Mona Lisa,* had yielded a high ratio of irony to success. The fact was, if Lascaux was perturbing ideas of modernity it was because the modern world had perturbed the cave far worse. In this contemporary battle of the books, the sciences commanded the heavy ordnance and horse. This time, however, its charge was not to destroy the ancients but to preserve them.

Still, the deterioration continued—a literal decomposition. Every effort to stabilize only resulted in another upset in what became a bizarre dialectic between past and present that segued into something like an arms race. For decades administrators had required visitors to disinfect their shoes with a formaldehyde foot wash that, in the end, killed off benign microbes that might have attacked the malignant fungi that had begun to propagate. An air-conditioning system installed to hold temperature and humidity constant somehow catalyzed an infection by a white fungus that spread from ground to walls, like a snowstorm in reverse. Specialists spread the floor with quicklime and coated the walls with cloth marinated with antibiotics and fungicides. To ensure that an exact copy of the imperiled paintings existed, a detailed record was made for computer simulation; the strong, steady lights needed for the replication evidently spawned a black fungus that now crept toward the threatened scenes. A team outfitted with special suits applied an ammonia-based solution to the spots, and in nine out of eleven sites they halted further spread.

And so the cave's recent history has unfolded, like a reversed origami, with each intervention sparking a new crisis in what was becoming almost a parody of reductionist thinking, in which process replaced purpose. In January 2008 the cave was shut even to researchers and staff. The chief administrator, Marie-Anne Sire, lamented that Lascaux was "in the hands of doctors who don't have the same diagnoses." Some argued for further intervention, some for none, while all recognized that the dangers of doing and not doing might be equally great. Meanwhile, Altamira barred visitors in 2002 and diverted inquisitive visitors to a replica akin to Lascaux's, until in 2010 the Ministry of Culture voted to reopen the cave on a limited basis.[24]

The Ideal Becomes the Real

Before Lascaux and other sites of Pleistocene art became known, the most famous cave in Western thought was Plato's. In *The Republic*, his depiction of an ideal state, he sketched an allegory of the human condition in which he laid out his understanding of the nature of reality. His thought experiment became the archetype of all later imagined caves. The Allegory of the Cave is Plato's word painting of what it means to be human.

We are, he insists, like the occupants of a cavern utterly removed from the light of day. There we sit, shackled by our material existence. What we see is what a hidden fire throws up on the walls and ceiling. We know only shadows. We discern only a flickering pseudoreality in that numinous region where unstable firelight throws forms onto rock. To know the real world, we must shatter our fetters, leave the cave, and experience the true light of the sun and the ideal world it illuminates.

The Cave of Lascaux has perhaps become an unintended realization of that allegory. Its scenes were created by firelight projected on stone walls and ceiling. The difference is that we have reversed the process. We must leave the nominally real world to visit them. We make their hard substance into ghostly ideals and, in so doing, we unsettle their permanence. We must then re-create them in simulacra, with faux firelight, all of which in turn become unstable such that what we find in the Pleistocene caves is a shadowy simulation of what the lights we kindle, be they Magdalenian or modern, can project. The images are both substance and shadow, and, if we are to understand them, we must accept their dual character, which, as Plato suggested, is our own. When contemporary humans exited Plato's cave, they walked not into unadulterated sunlight but into the manufactured glare of Lascaux II.

PART THREE

HOW THE PLEISTOCENE LOST ITS TALE

We sow corn, we plant trees, we fertilize the soil by irrigation, we dam the rivers and direct them where we want. In short, by means of our hands we try to create as it were a second nature within the natural world.

—Cicero, *De natura deorum* (45 BCE)

By their fruits ye shall know them, not by their roots.

—William James, *The Varieties of Religious Experience* (1902)

Some 22,000 years ago the last glacial reached its maximum and then shut down, as had other periods of glaciation throughout the Pleistocene. With startling rapidity, broken by an occasional spike of cold, the ice began to melt. As before, it lingered in protected patches—some large such as Greenland, Antarctica, and the Arctic Ocean, but most small, nestled on the summits of mountains and in niche valleys—and it vanished at various times. But a consensus places the general collapse at 11,500 to 14,500 years ago. Those dates mark the onset of the contemporary interglacial. By convention they also delineate the upper border of the Pleistocene and announce the onset of the successor epoch, the Holocene.

A certain logical uneasiness patrols that border, since the defining dialectic between the geographic and genomic narratives has not ended. An interglacial is just that: a pause between glacials. The cosmological rhythms that have originated and governed the Pleistocene glaciations have not stopped; the ice will return. So, too, modern humanity has evolved continuously and has flourished, with cadences that are the cultural equivalent of glacials and interglacials; there is no clear demarcation that might justify sealing off its history. Concluding the Pleistocene with the advent of the current interglacial is, by the usual criteria of the geological timescale, arbitrary. It would seem an act of whim, narcissism, or hubris to segregate the period of modern human prominence from the grand cadences that have shaped Earth for 2.6 million years.

Yet if the Pleistocene is to end, it must end somewhere, and all the designated periods of the geologic timescale are arbitrary, in that they reflect the interaction of the human mind on material evidence. What makes its successor epoch, the Holocene, distinctive is that the human mind and hand have begun tampering with the evidence itself, steadily overwhelming the Earth system. This time the grand matrix has not been reloaded so much as reinvented, or perhaps dissolved. The Earth system no longer derives solely from geophysical inputs like Milankovitch cycles or genomic threads, so much as it reflects what Cicero long ago characterized as a humanly refashioned second nature. But in disrupting and reorganizing the interactions of earth, water, air, and fire, modern humanity has also created the need for a new narrative.

Modern humans have changed the fundamentals. They are less and less governed by the bridges and barriers initiated by climate, sea level, ice, and sand, and by competition with other megafauna, and more and more by the cultural world—symbolic, moral—that humanity has created for itself. Their inherited genome is not a foreordained future that they must live with but something they can tinker with. A new master narrative is taking shape in which geography and genome do not drive humanity so much as humanity drives them. In the past the major transitions and migrations of Pleistocene hominins map well against prominent gyrations of climate and the shakeouts it created. But steadily humanity has acquired greater ability to shape its surroundings, itself, and its understanding of how it interacted with geography and genome, until the time has come when it has acquired the capacity to upset and perhaps supplant the byways and workings of Earth's atmosphere and biosphere. It remains unclear just when that transition occurred, or how rapidly, or by what means. But the outcome is not in doubt.

What had been immovable now moves. Modern humans have become themselves the prime movers. They have rendered the past no longer a predictable prologue to the future. More and more humanity sees only itself—and seeks to explain itself and its surroundings—in the looking-glass world it has created around it.

Chapter 9

The Hominin Who Would Be King

[M]an is, in both kind and degree, a power of a higher order than any of the other forms of animated life . . . [and] modern ambition aspires to yet grander achievements in the conquest of physical nature.

—George Perkins Marsh, *Man and Nature* (1864)

IDENTIFYING AN END to the Pleistocene has proved as vexing as finding a beginning. Over and again the sense has prevailed that modern humanity deserves its own epoch on Earth. Lyell spun off the Recent as that period "tenanted by man," which in 1833 he thought would extend a bit beyond the age of ruins and written records. By the 1860s it was clear that the "antiquity of man" was far older. Yet the sense endured that modern humanity warranted its own nook in the geologic timescale. The ice was gone or going, the human presence on the planet was magnifying, and the Pleistocene seemed to belong to a former world. That epoch could only cease, however, if it had a formal closing. What Lyell called the Recent and others replaced with Holocene served that purpose. It identified the closing of the Pleistocene with the opening of the most recent interglacial.

Yet like many indices of the Pleistocene, this one has proved unstable. The cycle of glaciations has not stopped—it has only entered an interglacial phase. Nor has hominin evolution ceased—it has merely shifted emphasis. By the criteria used to establish the Pleistocene the epoch has not really concluded. The ending of the Pleistocene may have even less validity than its beginning. The Holocene hovers on the verge of extinction as much as California condors, Assam rhinos, and other lingering relics of the last lost world.

Yet an ending is needed. Today only one hominin, *Homo sapiens*, is left standing from the Pleistocene milieu, and the recent history of Earth is a story that sapiens are not merely the product of, but the producer. He is the intrepid hominid who, as Rudyard Kipling might have put it, would be king. The Great Game, which plays out across the Pleistocene and into the Holocene, ends with a trifecta of narratives, all of which replace the physical environment with humanity as an organizing principle. It thus shifts the question of cause from nature to culture, the determination of analysis from natural science to other scholarships, and the burden of judgment from mechanisms to motives. Humanity has not only remade Earth into its abode but has bonded it to a moral universe. The three braided narratives show why the world that made ours will end with the world we are making to replace it.

The Geographic Narrative: The Great Realignment

The Ice Age whipsawed species into and out of existence. But it ended by segueing into one of the great eras of extinction in Earth history. The episode reaches from the last glaciation to the present; this time, however, the interglacial did not restore variants of the lost species but continued, first wiping out megafauna, and then flora and fauna generally. Like a long oceanic wave that steepens into a tsunami as it approaches land, the wave of extinctions has quickened as it nears the present.

The Biotic Transition

The extinction question may be the sharpest example of both the clarity and the ambiguity of the larger biotic transition. Species had arisen and disappeared throughout the Pleistocene; even without the

Holocene's contributions the epoch could claim standing as one of
Earth's most prominent periods of extinction. But prehistoric and
historic evidence abounds of modern humans entering new lands and
extirpating their larger species—those with which it competed. On
Mediterranean islands, Pacific islands, and Indian Ocean islands pygmy
elephants, moas, and dodos all melted away rapidly with the advance
of people. Some were killed outright; they were clubbed, speared, or
shot in such numbers that the population fell beyond the point from
which it could recover. Others succumbed to humanity's fellow travelers, like rats, foxes, and hogs, that could burst into an ecosystem
with feral lethality. Still others ceased to flourish as humans and their
domesticated species restructured landscapes with fire, ax, hoof, and
weeds. Whatever the matrix of methods, the outcome was a torrent of
extinctions. Between 3,000 and 50,000 years ago, some two-thirds
of mammal genera and half of all species that weighed more than
roughly one hundred pounds went missing. The rates of extinction
are estimated at one hundred to one thousand times that prior to the
time of human dominance.[1]

The shock of lost species was greatest in those lands most removed
from the hominins' African hearth: Australia and South America lost
the most, and North America somewhat less, while Africa shed the
least. This asymmetry argues for the extinguishing power of hominin
colonization. Interestingly, when humans returned to Africa, as they
did in the late nineteenth century (outfitted with new weapons), they
set in motion another plague of megafaunal extinctions until forced
to stop. Clearly, at some point modern humans ceased to be among
the driven and became the driver.

Extinction was not the only outcome. Even as large species were
pushed to oblivion, humans began to capture, tame, and ultimately
domesticate select species. Artificial selection replaced natural selection, and the lost diversity of wild species is somewhat compensated
for by varieties of servant species. The roster of domesticated, or
tamed, mammals alone is astounding: dog, sheep, goat, cattle, pig,
horse, ass, cat, elephant, camel, llama, reindeer, rabbit, ferret, hamster,

guinea pig, and assorted lab animals. There are even speculations about tamed cheetahs and dolphins. Moreover, each species shows selection for many variants or subspecies. Some two hundred breeds of sheep exist, for example, a few of which may have gone feral and appear now as "wild." There are between two hundred and four hundred breeds of dog, and more if recognized hybrids are counted. The breeds and types of horses and ponies number in the hundreds. These creatures have extended the reach of humanity and emphasize the post-Pleistocene paradox in which there is more mammalian biomass now than before but concentrated into fewer species—and those under the rein of humans.[2]

The story of megafaunal mammals is thus the story of their relationship to modern humanity. They were extirpated as competitors or fodder, or they were domesticated, or they were actively protected. It is estimated that at the conclusion of the Pleistocene the growth of human biomass matched that of the vanishing megafauna. Then more species disappeared and humans began to multiply, both processes that went exponential over the past few centuries, such that most of the mammalian biomass in the world today is either human (40 megatons of carbon) or human domesticates (100 to 120 megatons of carbon), while wild vertebrates claim only 5 megatons. The natural biodiversity of the world has been replaced by cultural diversifying through humanity.[3]

The Pyric Transition

Mammalian biomass is only a convenient index, however. One species did not simply replace others, as multiple copies of a single book might be swapped for others on a shelf. The process involved restructuring the entire library, from its energy dynamics to its cycling of chemicals to the interaction and migrations among remaining species. The whole biosphere, in brief, began to realign. Holocene environments have become "human-distorted relics of a once natural world."[4]

Begin with energy, for reformations on this scale depend not just on cleverness or tools but on power, which derived from humanity's species monopoly over fire. Aboriginal fire depended on the capacity of the landscape to propagate it. People overcame this limitation by cutting, draining, drying, and otherwise mincing up organics, and so altered the pyric character of their surroundings. They could burn more, and burn at times and places outside natural parameters, although only as much as they could coax or cajole from the living world. Then people discovered geologic landscapes, long buried and lithified, that they could exhume and combust; they could burn such fuels from the geologic past and release them into a geologic future. The first escalation in firepower was associated with agriculture; the second, with industry. Humanity acquired almost unbounded sources of energy.[5]

The scales and rates of anthropogenic change speeded up until modern humans were redirecting even the major biogeochemical cycles of Earth. Carbon, nitrogen, sulfur, phosphorus, and metals like lead and iron—human ingenuity, meddling, and power tapped into their flows and redirected them. Using fossil biomass as both energy and raw material, people created new substances, some benign, some toxic, some degradable, some destined for biochemical immortality. They have even replumbed the hydrologic cycle. Some 36,000 dams existed by the end of the twentieth century, and two-thirds of planetary rivers were regulated; human usufruct claimed over half the global runoff of fresh water, rivers were rechanneled and siphoned into city and field as well as drained, and some inland seas ceased to exist. Between 10 percent and 40 percent of all net primary productivity has come under the direct or indirect auspices of humanity.[6]

Earth, air, water, fire—every element was reorganized, or chronically disorganized. The biosphere: extinctions, invasions, the relocation of species, the rewiring of energy flows and the jigging of biogeochemical cycles unmade and remade ecosystems. As with species, so with habitats; they collapsed, became domesticated as agriculture or cityscapes, or found formal protection as preserves. The atmosphere: the effluents of industrial combustion and the by-products of chemical syntheses polluted Earth's envelope of air, causing greenhouse gases to build up, ozone to deteriorate, aerosols to linger, and precipitation to become acidic. The hydrosphere: oceans warmed, changed chemistry, and threatened to reroute currents; artificial reservoirs flourished and natural lakes shriveled; rivers were dammed and diverted; and the pattern of rainfall changed along with the pathways of storm tracks and the frequency of intense events. The pyrosphere: two realms of combustion now define Earth, with the open burning of surface biomass but a fraction (40 percent) of the burning of fossil biomass. As a result, developing nations tend to have too much of the wrong kind of surface fire and industrialized nations too little of the right kind. But the sum is worse than these parts, because each sphere intermingles with the others.

All this is the standard fare of what has come to be called global change. So vast is the scale, so rapid the turnover, so banal is the expectation of more that a significant fraction of human inquiry is now directed at measuring the magnitude and character of the Earth that results. What inevitably emerges from such studies is the commanding, even unifying, agency of humanity. People are the single common feature that connects all the parts. The Pleistocene is the world that made ours because it made us.

The Climatic Transition

Inevitably, planetary reconstructions on this order must affect climate. What astonishes, however, is that these reconstructions do not merely modulate the effects of climatic forces—do not simply force the processes of climate to run through novel channels—but that they have begun to alter the fundamentals that create those processes. Increasingly the climate is no longer a presence outside human control. What had been a prime mover, in the Aristotelian sense, has become an epiphenomenon. Instead of declining into denouement the narrative arc shot up and the span lost its anchor.

As with so many measures of global change, there are serious uncertainties about when the shift occurred with sufficient power to be meaningful. There was clearly a time when the climate operated autonomously from any human wish or fiddling, and there has come a time when the effects of human practices are clearly perturbing the dynamics of climate as fully as they have biotas and landscapes. Because the Pleistocene is defined climatically, by its ice age, a determination of when anthropogenic effects became pronounced can affect the placement of its upper boundary as much as identifying the onset of glaciation has its lower.

The last glacial reached its maximum around 22,000 years ago, and the current interglacial commenced around 12,000 years ago. The Pleistocene has had far more time under ice than not, perhaps 80 percent. The best reckoning is that over the last 800,000 years interglacials have lasted around half a precessional cycle, or 11,000 years. Of course anomalies exist. For obscure reasons there was one around 400,000 years ago that seems out of sync with Milankovitch rhythms, and it lasted longer than others. But commonalities are greater, and predictions based on past episodes suggest that Earth should be heading toward a new glaciation. As paleoclimatologist William Ruddiman emphasizes, "The next glaciation is not 'imminent'; it is overdue." Instead of cooling, however, the planet seems to be warming at unprecedented rates.[7]

The models that have proved so effective in decoding the rhythms of Pleistocene glaciation predict that a new glacial should have begun in high latitudes roughly 5,000 years ago. The Little Ice Age that chilled Europe—variously dated from 1250 to 1900 or 1550 to 1850—should, according to past precedents, have snowballed into a full-blown ice age. But the Arctic ice didn't materialize; it continued to ablate such that the Arctic Ocean may have its ice burned away within a decade. Nor did the Little Ice Age swell into a new Fenno-Scandinavian ice sheet; it yielded instead to a period of moderate cooling and overall warming. The long summer, as Brian Fagan has termed the interglacial, persisted and segued not into a cosmological autumn but an anthropogenic spring. The fear presently is that warming may be irreversible, that in place of a Little Ice Age we have created a Little Scorch Age and Earth may return to temperatures it hasn't known since the Paleocene-Eocene (56 million years ago) or Toarcian (180 million years ago). The stable, favorable climate that allowed modern humans to flourish when released from the rigors of the Pleistocene may turn viral and become as burdensome as ice. We may be going from one extinguishing extreme to another, from the Big Chill to the Big Heat.[8]

The forecast trends for declining carbon dioxide and methane, extrapolated from past events, have broken down. Carbon dioxide veered away from the norm some 8,000 years ago and methane around 5,000. Both, moreover, are not merely declining at slower rates, or even holding steady, but are escalating in ways that cannot be explained by appeal to the usual suspects, from solar insolation to aerosol screening to unsettled biogeochemical fluxes. The most likely novel factor is a modern humanity that grasped more of Earth through agriculture. By slashing, draining, burning, and loosing domesticates on reconstituted landscapes, people have released stored carbon and stimulated methane production. The liberated greenhouse gases have created a greenhouse effect that has steadily made Earth into a self-steeping Crock-Pot.[9]

It is always easier to break than to build. Humanity does not now "control" the climate. It has shattered the old system of regulations that Earth had known through the end of the Pleistocene and has no real idea how to replace it. That controls are shuffling from natural agencies to anthropogenic ones is little doubted now. Most analysts, moreover, are more concerned with predicting the future than reconstructing the past. But just when the transition began, and how, matters for Pleistocene studies because it suggests an upper boundary for dating and reflects on the character of modern hominins—the twin themes that have traditionally defined the epoch. The old narrative had relied on an unmoved mover, climate, which has reincarnated now as modern humans, who turn out to be an unreliable narrator. The old narrative arc is no longer something that resides outside people but rather something they shape.

The Genomic Narrative: The Great Himself

The human genome is also changing, and again, it is changing not solely due to outside forces and natural selection but from choices made, consciously or not, by people. The erectines propagated wherever natural conditions allowed; the sapiens, wherever they could connive to get to.

Cultural Selections

The hominin takeover of Earth has happened, by geologic standards, in the blink of an eye. Having once firmly departed Africa it took modern humans about 40,000 to 50,000 years to migrate to all the inhabitable continents. That task was completed by the time the Pleistocene had inflected into the Holocene. It then continued, as people tracked the still-receding ice and as they took to boats to colonize islands, and it didn't end until they established permanent bases in Antarctica. What made it possible was that it happened in defiance of the kinds of geographic checks and climatic counterbalances that had confined hominins throughout the Pleistocene. The genomic narrative was no longer honed against the geographic, like the two shears of a scissors. The dialectic in which genes proposed and the environment disposed broke down. Selection apparently followed another logic.[10]

Of course natural controls did not themselves vanish. There is plenty of genetic variation even among modern humans, and hence room for natural selection to operate. Areas rife with malaria promote genes that also favor anemia (sickle-cell in Africa, thalassemia in the Mediterranean). There are genes that promote lactose tolerance, that allow certain starches and sugars to be stored in times of scarcity, that lighten the skin to better absorb vitamin D, and that adjust body size, and the genes behind these traits have been promoted or discarded as they suit environments or as they follow the fashions of sexual selection. They continue today.[11]

What allowed humanity to swarm was not that nature withdrew but that humanity's capacity to block or cover natural checks became so large that it swamped natural effects. This is an observation that traces back to the founders of the field. Almost as soon as Darwin and Wallace had identified natural selection, the argument emerged that cultural forces were replacing natural ones. More and more, who flourishes or fades, who survives to reproduce or who doesn't, is decided by social circumstances, not by the implacable workings of natural forces. Society intervenes between individuals and nature, and culture intervenes to shape society. Simple surgery can correct the otherwise fatal genetic disorder pyloric stenosis; insulin injections can be used to manage diabetes; social institutions can help Down's syndrome babies to thrive.

Yet all such means are indirect in that they act on the phenotype. They alter the environment within which the genotype must operate; they swap one set of selective pressures, set by nature, for another, set by culture. Increasingly, however, those influences are moving toward the genome itself. If survival requires success in a cultural setting, understood largely, then selection will shift the genome in that direction. Moreover, genes express themselves within a cellular setting—an epigenome—that affects what genes are actually activated, and this process of epigenesis takes its cues from stimuli such as hormones that are in turn triggered by a being's physical surroundings. The effect has been documented for imprinting, embryonic development, metabolic networking, and gene activation for certain cancers.[12]

The epigenome intervenes between the outside environment and the genome proper. As people progressively shape that outer habitat, they effectively preselect within the inherited genome. A Pleistocene hunter will know one environment; a student immersed in iPods, laptops, BlackBerrys, and other digitech will know another; even with the same genome the two people may grow into different phenotypes. The traits necessary to follow a wounded mammoth are different from those needed to multitask. The matrix within which modern humans live is less one of earth, water, air, and fire than of silicon, synthetic foods and drugs, self-constructed and self-contained *domi*, and electrical fire from dynamos. As people have shaped dogs and potatoes to suit their needs, so they are increasingly shaping themselves.

It is only a matter of time before direct intervention replaces indirect. Genomic technology is escalating. In 2000 the sapient genome was sequenced, all three billion base pairs; in 2010, so was the Neanderthal. By then the genomes for some forty other eukaryotes and nearly twelve hundred prokaryotes were known, a bacterium had been created from a synthetic genome, and genetically modified foods were common. DNA testing had become a staple of forensics, an updated version of fingerprinting. Although early experiments faltered, DNA replacements will supplement organ replacements, while stem cell technology and lab cultivation of tissues may remake the procurement of organs and outright cloning will be feasible. Ultimately, humanity may be able to design itself free from natural selection. In short, the genome as a hard index against which to measure nature and as an axis of narrative is becoming biotic putty.

The belief endures that people need to be better than their inherited nature allows. Typically it has taken spiritual forms, the need to overcome what St. Paul called "the natural man," and it can veer into the quixotic, that in order to save itself humanity must abandon its bequeathed self for an improved version. Now, however, transcending the natural man can take material forms, with an infusion of grace through paired amino acids and the flow of electrons through microprocessors. More and more, overt and widespread interventions will become inevitable. They will likely begin with humanitarian goals, such as correcting genetically inherited diseases, and then go where money and power take them. In other words, people have begun changing what had always been, in their experience, unchangeable.

Mechanical Mind

There is yet another dimension emerging. From the earliest musings about what makes humanity what it is, sentiments have returned to the question of our self-awareness, our intelligence, our capacity to reason, abstract, and invent; in brief, our mind, or its physical manifestation, the brain. A big brain is shorthand for what most distinguishes us from other creatures and preceding hominins: it's what makes culture possible. Today, that conception is being challenged, in part by research with other creatures who also display intelligence and in part by information technology.

In one sense, information technologies simply do for brainpower what levers and power tools do for muscles. They extend, speed up, and magnify abilities to process data and act on that information. Instructed by people they can manipulate symbols, and even create new outcomes. No longer can modern humans assume that they are the only entities that can do algebra, play chess, or compete on *Jeopardy!*; not only can machines do such activities, they can do them better. Though the claims of researchers in artificial intelligence— the visionary beliefs that machines will be able to think and choose in ways not predestined by their programming—may be grandiloquent, they are challenging notions of intelligence as an index of human uniqueness, and they are further distancing humanity from the natural environment that had previously sustained and controlled it. In fact, as environmental scientist Vaclav Smil notes, "[w]henever it comes, if it comes, such a new form of machine-based consciousness would not need the Earth's biosphere for its functioning."[13]

To date, this has not happened, and like the prospects for controlled fusion, artificial intelligence may remain forever a technology of the future. But whether machines will ever think like people, people are certainly capable of thinking like the machines they make, and this may be the true upshot of the experiment. Much as people are tinkering with their genome, so they are also, in effect, tinkering with their capacity to reason. Over millions of years evolution had selected for a particular kind of intelligence that enhanced survival amid a world of competing creatures, who were also the product of natural causes. The development of information technology, whether or not it leads to artificial intelligence, is creating an alternative context for thinking—niche construction in virtual reality. What "culture" does to shield human "nature," technologies for the mind are doing for the human brain.[14]

Apparently it is not enough for modern humans to displace natural forces with the self. They must, it seems, meddle with the internal workings of that self. Humanity has become a planetary force. Where it once had the capacity to name geologic epochs, it now has the ability to deform them, or in the case of the Pleistocene, further unmoor an epoch from its anchor points.

The Cultural Narrative: The Great What-If

Those decisions will be made on the basis of how humanity chooses to define itself. They will follow—with ample allowance for serendipity, accident, and unexpected consequences—from values, ideas, beliefs, and desires. They will make sense and be judged by the cultural world that humanity inhabits, or, to paraphrase William James, by what the temperament of the times feels is right. The Pleistocene will become a moral world no less than a geologic epoch.

Nature's Pleistocene

This has always been a vital attribute of the Pleistocene as an idea. From Agassiz's epiphanous vision to Buffon's meditation over the vanished mammoth, the last lost world spoke to the kind of Earth humanity inhabited and to the kind of character and agency humanity needed in order to live in that world. Accordingly it had an element of utopianism—the world as it ought to be.

The original imagining of the Ice Age conjured up a brutal, violent landscape of giants, dangers, vanquished beasts, and a struggling humanity that seemed the very embodiment of Hobbesian horrors: a life nasty, poor, brutish, and short. The Pleistocene provided a baseline for how far the human species had come—revealed the power of civilization, testified to the ennoblement of spirit over raw strength. So, too, it served as a cautionary tale about where humanity would fall if it failed in its civilizing mission, not so much just-so as what-if stories. From the legend of the feral boy to the hulking cynicism of Jack London's Wolf Larsen, atavism meant a return to the amoral spectacle of the Pleistocene with its unyielding ice and menacing creatures. Both visions understood that, while nature might change its appearances, might spin in and out of glacial epochs, and might invent species and take them away, it was immutable in its fundamentals that while its processes might mix in unpredictable ways, they were themselves invariant. Change might be a condition of nature, but nature did not itself change in its essences. Morality could appeal to natural law for guidance.

Predictably, such convictions led to counterconvictions in which the lost world became a nobler one. They could relocate Montaigne's noble savage from Brazilian cannibalism to the painted caves of the Pyrenees. They might reincarnate Rousseau's state of nature into an interglacial that ice had scrubbed of encrusting social obligations, courtier classes, and Versailles. But efforts to envision them as sturdier and purer progenitors was a niche enterprise qualified by the self-evident need to improve the human condition, both materially and mentally, and by the still-overweening presence of nature as a shaper of human life and meaning. By the late twentieth century the more industrially advanced nations saw themselves and their relationship to nature differently.

The essences that once appeared immutable no longer seem eternal, and with that disillusionment has passed the natural basis for a morality beyond human invention. Of course such considerations leave unaffected religious creeds derived from sources outside material nature, but even most religions look to nature for confirmation. The very success of civilization in tinkering with climates, biotas, and genomes, however, makes any such corroboration problematic, and with it, an ethics justified by appeal to a natural order. It dismantles the tenets of a natural morality as Darwinism did William Paley's natural theology. "We can no longer imagine," as commentator Bill McKibben put it, "that we are part of something larger than ourselves— that is what all this boils down to."[15]

Utopian Pleistocene

Unsurprisingly, the resulting angst has sparked a reaction, not least among secularists who had swapped one creation story grounded in revealed Scripture for another revealed in the Book of Nature. They are less likely to celebrate civilization than to denounce its discontents; instead of progress, they see decline. They have come to imagine the Pleistocene less as a cautionary tale of potential degeneration than as the marker by which to measure a proven declension in the standard of human goodness and happiness. The newly imagined Pleistocene is, rather, an exemplar for what Earth—and its dominant species—might become. The Pleistocene as a moral narrative has gone from being a feared purgatory to a prospective utopia. The modern Pleistoceneans have seen the past and it works.

The power of the vision is most pronounced among those who look to a natural order (an updated version of natural law) for guidance. These are mostly environmental thinkers of one sort or another for whom the Pleistocene was a golden age in which people and Earth coexisted and humanity remained within its naturally prescribed limits. That era ended with a broken bond between nature and culture, and several visionaries have proposed measures to correct it. One outcome is to create, as relics of that past, pristine preserves from which modern humans are barred (a kind of Eden in reverse). Another response, given the absence of suitable vestiges, is to re-create the Pleistocene here and there as a kind of natural theme park. Some advocates would like to rewild cultivated countrysides, reversing the process of civilizing and returning those landscapes piece by piece to what they were at the creation. Unwittingly, perhaps, they echo Bertrand Russell's observation that when you trace the expression "state of nature" to its source, it means the world the author knew as a child. The Pleistocene as humanity's youth—here is a Pleistocene childhood's origin to Arthur C. Clarke's futuristic vision of Childhood's End.[16]

Many point out that the Pleistocene imprinted itself on our behavior, our genome, and our temperament, and the further we move from those founding criteria the more we invite diseases, dysfunctional societies, and damaged psyches. The chronic diseases of the present, such as type II diabetes, are the result of living out of sync with our formative experiences. The reckless wastage of natural resources stems from a creature accustomed to feast-and-famine cycles and unable to live sustainably in a world without clear limits instead of the (momentarily) unlimited world of its own making. The breakdown of social order is understandable, because men and women carry within them hardwired instructions for behavior encoded in a past epoch that bear little relation to the places they now inhabit. Such reasoning applies to the mind of modern humanity no less than to its body. The Pleistocene, it is argued, shaped our collective cultural genome, and we are barely able to understand the post-Pleistocene world we have conjured up. Responses point to two poles. Either you tinker more furiously to correct those maladaptions, or you cease and let the past reclaim its future.[17]

From its first conception the Pleistocene as an idea has held lessons about how nature worked—experiences that translated not only into a better scientific description of Earth but of how people ought to live on that Earth. Now the comparison is more subtle, disjunctive, and trickier. The modern world that measured itself against the past appears to operate on very different principles, for which the Pleistocene is no longer a guide. In fact, the creature responsible for restructuring that world has projected those uncertainties into the past and has destabilized its definition. One variable now chases another as morality twists into a Möbius strip.

Chapter 10

The Anthropocene

Man, one harmonious soul of many a soul,
Whose nature is its own divine control,
Where all things flow to all, as rivers to the sea.

—Percy Bysshe Shelley, *Prometheus Unbound* (1820)

No one ever went broke overestimating people's sense of self-importance. Humanity consistently finds itself to be its own most interesting theme, and consequently functions with an exaggerated view of its significance. Celebrants exalt that Earth was created to be humanity's habitat, while critics rail against the presumption that Earth's largesse exists only to serve people who have ill-served the planet, and moralists denounce humanity's self-absorption.

Yet the magnitude of global change in recent centuries has been so extraordinary and so linked to anthropogenic activities that the human presence has become an undeniable agent acting on a planetary scale. What had been handprints on the walls of some Pyrenees caves have become hands on the levers that govern the world cave. So immense are the cumulative effects that modern humans have begun crediting themselves with the creation of a new geologic epoch for which their own actions, now challenging or surpassing those of nature, are the marker. The human presence is undeniable; it remains only for people to name it, and probably to name it after themselves. Such an epoch would establish the upper border of the Holocene. The proposed name is the Anthropocene.[1]

If accepted, the Anthropocene would frame the Holocene, which would become that period since the onset of the last interglacial until the late eighteenth century. The Holocene would acquire a closed chronology, for which one could in principle construct a distinctive narrative. But it would be hard to justify a new epoch on the order of 12,000 years. Either the Holocene should be absorbed into the Pleistocene, or it could bond with the Anthropocene to create a new epoch. Either outcome would unsettle the Pleistocene's upper boundary, because the debate over the reality of an Anthropocene is really about the end of the Pleistocene, since it marks a period when the informing principles traditionally used to distinguish that epoch, climate and human evolution, collapse into a single agency, ourselves.

Tellingly, the proposed dates of origin correspond to the times when, in its European hearth, Percy Bysshe Shelley wrote *Prometheus Unbound* and Goethe, *Faust*, whose prelude urges

> Use both the great
> and lesser heavenly light,—
> Squander the stars in any number,
> Beasts, birds, trees, rocks, and all such lumber,
> Fire, water, darkness, Day and Night!

It is an eerie anticipation of what, materially, has followed.[2]

Humanity's Planetary Presence

The case for the Anthropocene is simple. The impact of humanity is no longer symbolic or suitably assessed against moral standards. People have become a geologic force on a scale and of a type that has been used to define temporal divisions in the geologic past. If an ice age can initiate the Pleistocene, industrialization can announce an Anthropocene.

Humanity has displayed the planetary power of an exogenous event, more like a meteoritic impact than something that emerged amid evolutionary checks and balances from indigenous materials, more akin to Chicxulub than a mutating plague like the Black Death, and it has left equivalent records that can be measured by the kinds of indices that have traditionally justified periodizing the geological timescale. There is the record of extinctions: observers suggest that Earth is undergoing its sixth great extinction, and since such events have been used to delineate stratigraphic borders such as the Cretaceous-Tertiary boundary (65 million years ago), it is appropriate to use it now. There is the record of oceans: the seas are rising and have become more acidic, both of which alter the location and type of sediments deposited (and dissolved). There is the record of the atmosphere: it has swelled with greenhouse gases and sulfate aerosols, leading simultaneously to planetary warming and global dimming. Both changes influence global climate, with its add-on effects for geologic processes. And there is a record of soil changes, from increased sedimentation to moved earth. Even land, it seems, shows the same suite of options as air, water, and life; it is destroyed, tamed, or deliberately spared. Add these together and they constitute exactly the kind of recognizably distinct events within strata that have guided the determination of past periods of geologic time.[3]

Time is relative; it is measured by events as space is by matter. One reason for the bias in a more finely calibrated timescale as it approaches recent times is that far more is preserved and known. The past couple of centuries have more data than the first couple billion years of Earth. So dense is the evidence of events in contemporary times that the Anthropocene is itself becoming subdivided, with what some authors call a Great Acceleration after the conclusion of World War II—an increase of an "exponential character." Most observers identify the onset of the new epoch with coal-burning industrialization, which, usefully, corresponds to James Watt's invention of the modern steam engine and a rise in CO_2 as captured in Greenland ice cores. For convenience, the Anthropocene typically dates from the year 1800.[4]

Ironically, the Anthropocene offers surer criteria than those used by the International Commission on Stratigraphy to identify older periods, epochs, and ages that depend on a Global Boundary Stratotype

Section and Point, that is, on a recognizable stratum and site. The reason is that the origin and unfolding of the epoch, while recorded in the geologic record, do not arise from geologic causes but from human ones. Unlike the Pleistocene, in which natural events define human history, in the Anthropocene human events delimit natural history. The choice of an originating date depends on human activities as judged by human researchers. While "concepts of the 'naturalness' of boundaries underlie . . . current debates on the positioning of the boundary of the Quaternary period," they are becoming irrelevant.[5]

The Great Acceleration, as Will Steffen, Paul Crutzen, and John McNeill note, "took place in an intellectual, cultural, political, and legal context in which the growing impacts upon the Earth System counted for very little in the calculations and decisions made in the world's ministries, boardrooms, laboratories, farmhouses, village huts, and, for that matter, bedrooms." So while the Anthropocene has scrambled the dynamics of previous eras, it did not result from natural causes but from cultural ones; and while its dates are embedded in the Earth system, they derive from anthropogenic determinations about suitable causes and markers.[6]

The Holocene as the Next Last Lost World

Even before the Anthropocene was proposed, the Holocene had become anomalous. As a committee of the Geological Society of London observed, at 2.6 million years the Quaternary—traditionally the era that embraced both the Pleistocene and the Holocene—is by an order of magnitude the shortest period in the geological timescale. The Holocene is the shortest epoch by two orders of magnitude; it is the only one of the Quaternary interglacials "to be accorded the status of an epoch"; and it is the only unit in the whole of the Phanerozoic whose base is defined in terms of numbers of years from the present (ten thousand radiocarbon years before 1950). The epoch has, as it were, two beginnings that must somehow meet at a moving middle.[7]

The invention of the epoch began in 1833 when Charles Lyell urged the label "Recent" to cover the period of known human presence." Strictly geologic data would yield to the scholarship of human history and prehistory. But Lyell's dates (again) proved wobbly, and his use of English instead of a Greek neologism also left colleagues unhappy. Accordingly, Holocene was proposed in 1867, formally submitted to the International Geological Congress (Bologna) in 1885, and endorsed by the Commission on Stratigraphic Nomenclature (U.S.) in 1969. The collapse of the last glacial maximum and the appearance of modern humans seemed criteria enough. The epoch is routinely identified as the "chronological framework for human history."[8]

Only twenty years prior to the 1885 congress had the aftershocks of evolution by natural selection placed humans under the rule of natural forces. But already there were countervoices to proclaim the power of humanity to improve or destroy the natural world. The Bologna Congress occurred at a time when Europe, North America, and Japan were industrializing, but when most observers credited agriculture and deforestation as humanity's major influence on Earth. This observation reaches back to antiquity—Plato had lamented the scalping of Greece's soils—as had its moral message, that the greatest threat to humanity was humanity itself. The Industrial Age put those old jeremiads through a series of updates. The eighteenth century identified a putative linkage between drought and deforestation; the nineteenth, especially through G. P. Marsh, documented civilizations that declined from self-inflicted injuries in the form of environmental abuse; and the twentieth, a post–World War II outpouring that matched, with a roughly twenty-year lag time, the rate and visibility of planetary changes. The Great Acceleration might describe with equal aptness the great era of environmentalist literature.

The proposal for an Anthropocene—lopping off the past two centuries—leaves the Holocene not only mislabeled, since it would no longer be the "whole" of recent times, as the Greek would suggest, but an epoch of a scant 12,000 years. The Pleistocene would be more than two thousand times as long, and while we surely know over two thousand times more about the Anthropocene, it seems ridiculous to leave to the Holocene either its name or its autonomy. Either it should be absorbed into the Pleistocene to embrace the full range of natural cause–dominated events or into the Anthropocene to extend the new epoch back to the point where human influence interrupted the planetary workings with a spoor or signature both specific in agency and global in scope.

The issue is not that the geochronological timescale changes—it always will. At issue is the reasoning behind the change and its consequence. The upper border of the Pleistocene has again become a moving horizon, as it has been since Lyell. It is unstable by reason of the geologic forces it parses, the motives behind those forces, and the means by which they are analyzed. Logically it should end when humanity usurps control over significant fractions of the Earth system, particularly climate. But as the drivers of change shift from natural to cultural, not only is Earth unsettled but so are the means by which people understand those changes. The Anthropocene requires other disciplines of scholarship than science alone.

Remaking by Head and Hand

A future commission might also include narrative itself among those pragmatic considerations, since moving the border changes the possible story lines. It replaces what had been a narrative arc from two beginning points into one with a beginning, middle, and end, and swaps out an ongoing, continually lengthening narrative with a closed one. Absorbing the Holocene allows that diminutive epoch to become the denouement of the Pleistocene, not its successor. It anchors the epoch more securely at both ends and grants the narrative better closure, one that encompasses almost all of the ice ages and the span of hominin evolution that remain under the control of natural selection, which is to say, it defines the epoch by forces outside human influence. Understanding the Pleistocene requires more than a credible narrative, of course, but until one exists discussions will wobble and bob as the intellectual equivalent of Milankovitch cycles tug and stretch them.

Nor will repositioning the frame resolve the larger cultural crisis of narrative. Just as the Great Acceleration was demonstrating the power of humanity, the intellectual classes became disillusioned with narrative as an explanatory device. They didn't like the existing stories, and rather than replacing them or rethinking what an environmental narrative might mean in these days, they tended to ignore, deconstruct, or scrap the idea of narrative as a means of analysis and synthesis, of general expression. But that controversy was in many respects an intramural one among literary theorists and philosophers, like art critics arguing over whether two-dimensional canvases could legitimately represent three-dimensional scenes. The pragmatic need was for a story. Narrative did not melt away; it simply morphed.[9]

The creation of an Anthropocene can quell some of the unresolved themes and troubling instabilities of the existing Pleistocene narrative. There would be no arbitrary ending with the onset of the latest in a long roster of interglacials; the rhythms would continue. So, too, the awkward question of behavioral modernity would be displaced until such time as modern humans acquired sufficient technological power to challenge and selectively overthrow natural processes on a geologically recognizable scale. Adjusting the borders helps sort through the troubled from the healthy aspects of narrative, like the practice during a financial crisis in which "bad" banks are invented to take on the toxic debts and bogus collateral and leave "good" banks to reestablish credit and credibility.

The man who coined the term Anthropocene, Paul Crutzen, atmospheric chemist and Nobel laureate, has argued for urgency of action. "[S]cientists and engineers face a daunting task during the Anthropocene era: to guide us towards environmentally sustainable management." Such reform will demand "appropriate human behavior at all levels" and may require "geo-engineering projects" on a planetary scale. Having begun in hubris, humanity cannot withdraw in the face of nemesis, but must, it seems, do more, perhaps in the hope that, like Faust, "He who strives on and lives to strive / Can earn redemption still."[10]

Revealingly, Paul Crutzen said nothing about other breeds of scholars. Yet while the problems are technical, they are even more dauntingly political and ethical. Determining and applying "appropriate human behavior" is not a scientific matter. Engineers should know that technology can enable but not advise; scientists, that science can advise but not decide. Rather, the choice must reside in those other sloppy forms of inquiry and action that lie outside their purview. While other scholarships lack the rigor of experimental science—they can't build machines or construct machinelike behavior, they can't prescribe how, for example, to seed the ocean with iron or spray the sky with aerosols—they can enrich our understanding of their human context and the fallible terms by which we describe our understanding. (Or they can, if they cease self-inflicted injuries and move from problematizing to problem solving.)[11]

One of the consequences of Darwinism was to transfer attention from spiritual motivations to creature comforts, to accept and build upon the animal origins of humanity. One of the outcomes of modern environmentalism may be to reinstate a morality not derived from natural causes. To date Earth systems have treated the humanities rather as Darwinism did theology; yet the need to address matters of human identity, ethics, and meaning point to the pertinence of such scholarship in an updated form. More than tinkering with temporal borders, the recognition of the Anthropocene might force a shift in the borders between disciplines. It might compel environmental studies to ponder, if not incorporate, the moral universe that humans inhabit into their realm of interest and to search out methods by which to study it. It is unlikely that humanities scholarship will act on existing Earth science systems as humanity itself does on Earth, but the frontier that has long divided them might shimmer and blur and prove less impermeable than the more rabid partisans would wish. Such a conceptual overhaul would not be the first, or likely the last, paradox to emerge from the ongoing contemplation of the Pleistocene.

Behaviorally Modern Disciplines

The span proposed for the Anthropocene describes exactly the period during which the Pleistocene as an epoch was identified and studied. But that time was also an extraordinary epoch not merely for natural history but for learning overall as scholarship broadened and quickened. The Pleistocene as an idea was one outcome.

Mostly, contemporary observers commented on the widening chasm between the humanities and sciences, not merely in the quantity and caliber of their sources but in their relationship to the culture. Science's more expansive promoters believed that, with time, the novum organum of empirical collection, experimental tests, and mathematical organization could subsume all forms of knowing, or at least all those worth having. That, at least, was the prophecy of the bolder proponents of the Enlightenment who, following Condorcet, could imagine an endless progress of knowledge. By the time Louis Agassiz announced his Eiszeit, to be intellectually modern was to accept science as the primary mode of inquiry.

That was the context in which Pleistocene was imagined. It was defined by scholarships of natural history, which eventually split off and specialized into geology and paleontology, and would soon include archaeology, which was slowly crystallizing out of the classics and history and would eventually announce that it, too, was a science. Other fields of study emerged and joined, from paleoecology to climatology to glaciology—a panoply of disciplines, but all of them self-identified as sciences at least in name and intent. Whatever their origins and comparative anatomy, as it were, they had experienced some spark of creativity that set them apart, like those modern sapiens who colonized the Paleolithic of Europe and garnished the walls of Cauvet and Altamira with aurochs and reindeer. They behaved differently from their predecessors, and in ways that justified calling them modern.

Yet even as the two cultures diverged, they shared some common assumptions about how the world worked and how it might be understood. In particular, history was becoming a great organizing theme.

To explain something was, by and large, to describe its historical development: to identify its origin, to let its informing principle unfold, and to have its climax serve as a conclusion. The Pleistocene is a fabulous expression of this intellectual era, for it was itself a historical period with a thematic beginning, middle, and end. But could history be a science? It could, if two conditions were met. One, researchers had to strip the record of its nonempirical and untestable attributes. If some feature was opaque to scientific inquiry, it did not exist for purposes of constructing a rational narrative. And two, the organizing narrative had to parse away contingency. If the story did not unfold in predictable ways, it lacked the kind of causality that made for science, as science was then understood. There were discernable laws of history as there were for mechanics.

In these assumptions the tropes of classic narrative and the methods of modern science found common cause. Secular origin stories, founded on the data of new sciences, could replace those of the humanities much as humanist narratives had superseded those of theology. Natural history did what Carl Becker once observed of the philosophers of the age, that they "demolished the Heavenly City of St. Augustine only to rebuild it with more up-to-date materials."[12] Still, narrative, too, changed, becoming more organic and complex, much as it understood Earth and evolution to do. Eventually historical accounts, like the era's swollen novels, came to resemble a temporal version of the Ptolemaic system of astronomy, full of eccentric equants and epicycles and other seemingly ad hoc tinkerings.

Then came the modernist revolution, passing like rolling thunder over field after field. It struck physics, mathematics, art, literature, and biology early; it penetrated those disciplines most allied to Pleistocene studies later. For earth science, archaeology, and paleoanthropology it did not reform the fundamentals until the 1960s. Throughout, the modernist persuasion broke the power of narrative. The modern synthesis in twentieth-century biology, neo-Darwinism, decoupled historicism from evolution. Natural selection acted on chance mutations; there was no predictability to evolution; what seemed to be progress was an illusion in the mind that saw directional patterns that did not exist in nature. Geology replaced its epic narrative with a jumbled, idiographic story of shifting tectonic plates. At the same time modernist literature began scrambling narrative and narrators, breaking the valence that had made narrative a universal mode of explanation. The loose cultural bond that had joined the sciences and the humanities split further. It seemed that disciplines, like the sapiens, might be anatomically modern without being behaviorally modern.

Crisis of Knowledge

The crisis struck philosophy at the same time it did the sciences, and, with special vehemence, the sciences of hominin origins. At the end of World War I Neanderthals and the erectines of Java were known, and Piltdown Man was still considered legitimate. Then came the challenges: the erectines of China, the australopithecines of Africa. The entire enterprise underwent a slow chrysalis, from which it did not emerge until the 1970s. But between 1925 and 1936, as paleoanthropology debated the reality and meaning of *Australopithecus africanus*, philosophy began a similar metamorphosis as it addressed a "crisis of knowledge" that saw philosophy split into two major lines. At the beginning it was still possible to imagine a mosaic of disciplines that could contribute to a common picture. At the end, philosophy, like the scholarship of hominins, had seized on modern science as both subject and model.

One camp looked to the natural sciences for inspiration and sought to found even language—even the language of mathematics— on symbolic logic. The analytical tradition, as it came to be called, accepted as positive knowledge only that which could be verified by science, and accepted as science only those inquiries that met criteria best exemplified by the "exact" rigor of physics. The "continental" tradition, by contrast, evolved along the lines of classic philosophizing, continuing to answer the fundamental questions with updated, typically neo-Kantian, reasoning. It recognized that, in Hume's earlier phrase, there were "Relations of Ideas" that are "discoverable by the mere operation of thought" and "Matters of Fact" that are the result of experience and contingency, and that both forms of understanding have their place.

As tension built there were efforts to turn the competing concepts into complements. One of the most spectacular concerned Ernst Cassirer's endeavor to update Hume's two forms of knowing into two scholarships, a "science of nature" and a "science of culture." The science of nature encompassed the exact sciences; the science of culture brought together fields that normally resided within the humanities, for which "science" meant knowing generally, not restrictively that kind of knowing characteristic of the modern exact sciences. Cassirer's quest to reconcile culminated in 1929 with *The Philosophy of Symbolic Forms* in which he explained his recasting of Hume and analyzed how reconciliation might come through his concept of an *Objekt*.[13]

Every Objekt, he explained, has three features: natural science, history, and psychology. By natural science Cassirer referred to its analytic, physical properties—those aspects of a thing that can be ascertained with a reproducible methodology. By history he referred to the life history or the historical contingency of context that surrounds an Objekt—the events that encase it and formulate its context. And by psychology he referred to the metaphysical meaning imbued in the Objekt or assigned to it—the cultural clout that gives it meaning and weight. In this way, the "physical, historical, and psychological concepts continually enter into the description of a cultural object."[14]

Cassirer had no hostility toward science. On the contrary, he regarded it as "the highest and most characteristic attainment of human culture." But he recognized its limits, and like Karl Popper distrusted its intrusion into areas in which it could not properly function. In particular, he insisted that it was "impossible to 'reduce' historical thought to the method of scientific thought," for "[h]ere we are not moving in a physical but in a symbolic universe." The "rules of semantics, not the laws of nature, are the general principles of historical thought. History is included in the field of hermeneutics, not in that of natural science." Like Cubist images visible from several perspectives, every Objekt possessed multiple, simultaneous interpretations.[15]

Cassirer distrusted not science but scientism, as Popper distrusted not history but historicism. The fallacy lay in trying to impose the methods of the exact sciences beyond their realm of competence. To demonstrate his point, Popper cited T. H. Huxley, who declaimed: "he must be a half-hearted philosopher who . . . doubts that science will sooner or later . . . become possessed of the law of evolution of organic forms—of the unvarying order of that great chain of causes and effects of which all organic forms, ancient and modern, are the links . . ." Rather, evolution of life on Earth, "or of human society," is "a unique historical process" whose description "is not a law, but only a singular historical statement." What mattered for modernity, however, was that a discipline be a science, and if history could not be scientific, then aspiring disciplines would abandon it along with the appeal to narrative that was history's natural medium.[16]

Ernst Cassirer was not alone in trying to hold the fissuring blocks of Western thought together, but he is among the best known, and he came to symbolize the rupture. Following a famous disputation with Martin Heidegger in 1929, the two polarities pulled apart, like slopes splitting under the pressure of an upwelling, one side holding with Heidegger and the other with the analytical tradition of philosopher Rudolf Carnap that would lead to logical positivism. The Cassirean synthesis, in which natural science, history, and psychological meaning both reinforced and kept each other in check rifted apart.[17]

The Great Acceleration

That was the postwar inheritance. The Great Acceleration that proponents of the Anthropocene have graphed as following the war had an intellectual as well an environmental dimension. It had a political counterpart in technocratic development and the reemergence of a global economy, but it also had its intellectual doppelgänger in a shift toward a more rigorous and exclusive commitment to science. Those fields most pertinent to the Pleistocene reconstituted themselves along lines that explicitly emulated the natural sciences.

More properly, practitioners in these fields tended to copy what they perceived to be the essence of those sciences. They read Thomas Kuhn's *The Structure of Scientific Revolutions*, from which they identified those features that they understood made disciplines behave like real sciences, which is to say, as disciplines that could cluster around a consensual paradigm. They interpreted Kuhn's book less as a paradigm of science's past than as prophecy, perhaps self-fulfilling, by which new sciences could emerge. They read the prevailing philosophies of science, now dominated in the Anglo-American world by logical positivism about how "real" science worked. In retrospect it could appear they wanted a means to strip away the conceptual clutter and cultural biases that had become, to their minds, a limitation (if not an embarrassment), and in order to effect those changes, they sought the sanction of philosophers of science. They wanted a scientific revolution or a legitimized revolution of ideas. They would render unto science what was science's, and what wasn't they would disregard. But they believed that much of what intellectuals had previously assumed lay outside the purview of the exact sciences could, in truth, be absorbed. As Lewis Binford intoned, "Science is a method or procedure that directly addresses itself to the evaluation of cultural forms." In brief, either hermeneutics could become a science, or it could disappear.[18]

Among the glaring shifts was the abandonment of narrative. The study of geology became process driven, shunning the range of its inherited preformed narratives, from cycles of erosion to the epic of Earth. Archaeology too turned "processual" in practice, positivistic in its philosophy, and paradigmatic in its perspective. For example, Binford thought a "science of the archaeological record" was possible, and he disagreed with Kuhn that paradigms changed for seemingly irrational reasons. Rather, he was "convinced that we can learn to encourage productive paradigm change through rational means." In the end the point was not just to recharter archaeology on the model of the exact sciences, as logical positivists sought to remake even language on the model of logic, but to call for a revolution, a "new perspective."[19]

They got it, although not exactly as they forecast. Practitioners remained Baconian in their passion for amassing data, and in all fields of archaeology they tamped down the more radical positivism into working models like "middle range theory." But if the revolution did not achieve its theoretical goal, it helped dismantle the old ways of thinking. The only evidence was hard empiricism; the only mode of explanation, the style of hypothesis testing that characterized the envied sciences. There was no equivalent accommodation for a working narrative that might function in the way a working hypothesis did. So, despite public denials, the new disciplines looked like positivism, talked like positivism, and walked like positivism.

Against such brash manifestos, Cassirerean syntheses seemed quaint, akin to the inherited grand narratives that had sought to connect the sherds and lithic dots found in the soil. These narratives were too progressive and too providential, both naive and bombastic. They seemed to resemble parables more than paradigms and appeared closer to Plato's fables than to contemporary philosophies of science. The old classics had no more pertinence for Pleistocene studies than the Book of Genesis, and the old narrative forms no more explanatory power than the story of the Noachian flood. Worse, unless stripped of epistemic authority, the old myths might return, like Olympian gods visiting Earth in human disguise.[20]

The push for better analysis of artifacts undoubtedly improved the ground-truthing of archaeology. But it came at the cost of delaminating the kind of synthesis Cassirer had urged, and it unbalanced the dynamics by which disciplines acted on and checked one another. Science could prove a more efficient form of reasoning but in the same sense that autocracy can be a more efficient form of governing; left unchallenged, it could seamlessly segue into scientism. Nor did the emphasis on process resolve the matter. Process could take a path toward pragmatism or to positivism. By choosing positivism the larger discipline confirmed its self-identity as a science (and gained access to the funding of science), but it cut itself off from other scholarship. Yet as philosopher of science Paul Feyerabend notes, "If we want to understand nature, if we want to master our physical surroundings, then we must use *all* ideas, *all* methods, and not just a small selection of them," for "everywhere science is enriched by unscientific methods and unscientific results." The "separation of science and non-science is not only artificial but also detrimental to the advancement of knowledge." Particularly during the 1960s and 1970s, the objects of the New Archaeology lost their claim to be Cassirerean Objekts.[21]

Epilogue

Rift Redux

What if . . . he was turned toward objects that were more real, and if . . . he were compelled to say what each of the passing objects was when it was pointed out to him? Don't you think he would be at a loss, and think that what he used to see was far truer than the objects now being pointed out to him?

—Plato, *The Republic* (ca. 380 BCE)

ENDS, BEGINNINGS, AND LINKS between them—a narrative must somehow return to its past even if it is to trek far into the future. For our survey of the Pleistocene this points to revisiting the Rift Valley of Africa, which offers both a place of origin and a material metaphor for the vantage points by which that origin might be interpreted.

Creation stories—and that is what the Pleistocene has become—require a place of origin as well as a time. For some stories the site is materially grounded; for others, it remains symbolic; and for most it is both. It is the place where people first emerge. For the Hopi Indians it is the *sipapu*, an orifice in the gorge of the Little Colorado River. For the ancient Egyptians it was the Nile River, the source of mud and water. For ancient Hebrews it was the garden "eastward in Eden," the source of the four great rivers. For those interested in human origins it is, as often as not, the African Rift Valley. The quest for human origins has other formative sites, of course, notably in South Africa, and breakthroughs have appeared in the Neander Valley, Java, Flores, and even Siberia. But the Great Rift has repeatedly drawn and rewarded researchers, and it seems to symbolize in its very terrain the evolutionary splitting that led to modern humans.

So, too, it appears to stand as a material metaphor for the old rupture between the natural sciences and the humanities that now face each other across a widening gorge of incomprehension. In this cultural rift, both sides do not seem to move equally. Rather, one side—the sciences—seems to sprint away from the other. It moves quickly, shedding its own past even as it reveals the past recorded in rock and bone. In contrast, the humanities stand like an old cliff, slowly receding from erosion, holding its fallen scree at its base. One can still usefully read Aristotle's *Poetics* for its insights into tragedy. No one would read his *Meteorologica* as an introduction to earthquakes.

The future of the past, it would seem, must go to the fast-paced sciences as they relentlessly widen the gulf. Yet it takes two cliffs to make a gorge, and it takes the perspective from both rims to capture the panorama of the Pleistocene.

The Fall

In the late 1950s an American playwright, Robert Ardrey, turned his mind to the problem of human origins and directed his eyes to Olduvai Gorge within the Rift Valley, which he regarded as "the Grand Canyon of Human Evolution." After L. S. B. Leakey's recent discovery of *Zinjanthropus*, soon succeeded by *Homo habilis*, the gorge became "the world's most important anthropological site." His chosen title for the book that resulted in 1961, *African Genesis*, spoke to the twin themes of our conception: "Not in innocence, and not in Asia, was mankind born." According to Ardrey, we emerged from Africa, we descended from Cain, the killer, and we still bear his mark.[1]

A literary man, the inheritor of a humanistic education, Ardrey set as his ambition to interpret to the general public the palpable results and, even more, the moral meanings wrought by the past thirty years of scientific inquiry into humanity and its animal origins. The upshot was a silent "revolution," a "new enlightenment," destined to unseat yet another of humanity's self-regarding conceits about its own uniqueness and importance. Central to that thesis was not simply the

Leakeys' discovery of a tool-wielding hominin but the debate sparked by Raymond Dart's earlier discoveries of a darker, ancestral australopithecine already infected with predatory bloodlust.

It was an odd book, to say the least, and a wildly popular one. Its author took the latest in scientific discoveries—those vetted by specialists and "authorities"—and reworked them into familiar tropes. Interestingly, the design of the book did not sketch a master narrative, which its implicitly evolutionary theme suggested, but a sequence of essays, each building on a particular theme. In this way it emulated the style of an argument or literary assay. Each chapter assumed a form somewhere between an Aesopian fable and a Ciceronian essay and climaxed in what a literary critic might easily recognize as a kind of epiphany. The outcome read like a collection of moral epistles in a style that was closer to St. Paul than to articles in the *American Journal of Physical Anthropology*. Ardrey's self-imposed task was to bring a dramatist's flair to a grim fact: "a philosophical bomb, a positive demonstration that the first recognizably human assertion had been the capacity for murder." That mattered, because "the problem of man's original nature imposes itself upon any human solution." St. Augustine couldn't have put it better.[2]

The book is many things, not least a talisman of its times, particularly the vision of humanity as hunter and of conflict as hardwired into the human psyche. War had been a continual backdrop for Ardrey, who was ten when World War I ended, thirty-one when World War II began, forty-two when the Korean War broke out, and fifty-seven when the United States piled into Vietnam, and the specter of nuclear annihilation from the Cold War is never far below the surface of his text (the Cuban missile crisis followed less than a year after the book was published). The book gave dramatic voice to the fear that war might be continual. Ardrey's insight was to craft a narrative that could conclude by confirming contemporary fears, because it began with homicidal violence from australopithecines. "Cain slew Abel," he concluded, and humanity is his cursed offspring.

The trick was that Ardrey appealed to the putative authority of paleoanthropology. *African Genesis* seemed to reconcile the data of natural science with the ethical and epistemological concerns of the humanities. If it stood on the authoritative cliff of science, it peered across the rift with the eyes of an older scholarship and art form. Whether or not one agreed with his premise, Ardrey appeared to have spanned the opposing cliffs of human intellection—had devised a way to bring the two realms of understanding together through a study of the sole creature who practiced them both.

Yet the rift's fissures were widening even as the book saw print. The announcement of *Zinjanthropus*, with which the book opens, quickly collapsed, and by the time critics completed the first round of reviews, the earliest artifacts of *Homo habilis* had been found, thus shifting attention from what the putatively lethal australopithecines bequeathed to what the emergent habilines actually had. That same segue became true for every scientific fact on which Ardrey based his meditations. Philosophy, aesthetics, and ethics could not be securely anchored in the shifting sands of an active science. Either one used rhetorical skills to publicize or popularize existing data—dramatize the story of their discovery—or one risked imagining an ethical world whose premises trembled before each new revelation. Like many commentators on human origins, Ardrey tried for both.

He might instead have looked to an older genre that used nature as a basis for social commentary. Literary naturalists can overcome the hazard of the latest science by basing their contemplations on what they have themselves seen. They record empirical facts in the form of personal observations, which are irrefutable, and leave their judgments on a firm foundation not likely to crumble before the newest datum from a lab. The parables, fables, and morality tales that spin out of those observations can survive centuries of new learning. Rhetoric can sharpen the text much as statistics can a broken fossil femur. But an appeal to "what science tells us" makes poor pilings on which to erect moral structures. They speak to different purposes.

That is why *African Genesis*, when Ardrey speaks of his own experiences and invites readers to do likewise, remains a compelling contemplation of what it means to be human. But when, like many popularizers, Ardrey trades what he himself has witnessed and understood of the human condition with the putatively revealed truths of science, the book stumbles, dates in often embarrassing ways, and eventually joins the scrap heap of discarded theories. By basing his thesis on an interpretation of *Australopithecus robustus*, Ardrey put his *African Genesis* into the same category as the original Genesis—a successful literary text, flawed for interpreting natural science.

Literature's genres and tropes are powerful and enduring, the legacy of aesthetic selection over a long evolution. They become as ephemeral as yesterday's lab results, however, if they derive their conclusions from the latest scientific press release. They serve as publicity, not philosophy. They edge, as Ardrey's did, into a literary positivism that conveys the illusion of truth, as though they are only giving more vivid expression to certifiable facts. The issue lies not with the intrinsic properties of the mode of expression. The problem comes from trying to reconcile the substance of one discipline with the style of another.

The humanities do not need the sciences to satisfy their task. They cannot overtly contradict them, of course, but they can finesse around them, and they can incorporate the history and philosophy of scientific inquiry into their narratives. What they cannot do is derive their understanding from that of science, like corollaries reasoned from first principles. They have to span the valley from the cliff they stand on, not from its opposing echo, even if that involves a leap of faith.

The Other Side of the River

The sciences, too, have their cautionary tales. In 1968–69 hominin bones and tools were discovered embedded in a volcanic tuff near the Koobi Fora Ridge east of Lake Turkana. The igneous rock invited analysis by radiometric methods, despite the likelihood of contamination and mixing by resedimentation. Samples were subsequently analyzed by FM Consultants, Ltd., which employed both conventional methods of potassium-argon dating and the more recent ^{40}Ar/^{39}Ar technique. The first yielded a range of dates; the second, a keener determination of 2.6 million years ago plus or minus 0.26 million years. If correct, that number would revolutionize the chronology for the origins of *Homo*.[3]

What followed became known as the "KBS Tuff controversy"— "something of a legend," as Roger Lewin has put it, that affected everyone in the field of early hominin studies "however marginally involved he or she may have been." It ran for more than a decade that, usefully, coincides with the era in which a philosophy of positivism was colonizing those sciences committed to the study of human origins. Similarly it happened at a prime site, halfway between Olduvai Gorge and Ethiopia, and a scene of interest to paleontology and paleoanthropology for decades. It offers a cameo for the continuing crisis of knowledge that underwrites the retellings of the Pleistocene.[4]

Inevitably, given sparse data and quirky dates, protests arose. Researchers who had worked in the Omo Basin—what was loosely termed "the opposite side of the river," actually across the Rift and north of Lake Turkana—objected, based on the evidence of fossil pigs and elephantids, which they insisted could not be older than two million years. The splits multiplied. Paleontology challenged physics, stratigraphic dating misaligned with radiometric dating. Groups associated with strong personalities and entrenched research traditions pulled apart. Soon still other methods for dating were brought to bear, including paleomagnetism. The fissures widened until, in Lewin's words, they "cleaved the professional community in two."[5]

By 1980, however, the rupture had quieted down. New data from fossils and crystals were subjected to tests both traditional and recent. Careful mapping identified the stratigraphic sequence and its outcrops. Testing and retesting by various labs circled around a consensus opinion of roughly 1.8 million years. If figures diverged from that mean they were dismissed as the result of experimental errors, tainted samples, and murky geology. While outsiders might agree that the incident at Koobi Fora demonstrated, as Lewin observed, "how very unscientific the process of scientific inquiry sometimes can be," practitioners, however, tended to draw a different lesson from this particular episode. To them it showed that science worked, that it was self-correcting, that as researchers involved more data, more rigor, and more scientific disciplines they would converge on truth. They isolated the problem (in their minds) to the inevitable uncertainties of new techniques in a rapidly advancing field. They consigned the KBS controversy to the anecdotal prehistory of the discipline.[6]

The process had indeed worked for dating the samples. As Lewin points out, "[A]t one extreme, the KBS affair concerned the esoterica of complex geochronology, which virtually no one understood."[7] Eventually the machinery of science reduced that complexity into manageable pieces and then analyzed the fugitive data points. It solved the question of when the rocks had formed, and hence when the associated bones and stones had been deposited. The affair had been bruising and far from an ideal of open-forum science, yet it had, after a stormy decade, worked as the prophets of a more rigorously scientific discipline had forecast.

But deciding the date did not resolve the deeper issue. "At the other extreme was its implication for the antiquity of *Homo*, upon which virtually everyone had an opinion."[8] That, in truth, was the formative question behind the whole enterprise; it was what motivated the culture to sponsor such investigations in the first place. Getting more and better dots did not specify how to connect them, at least not in ways that could similarly resolve the formative questions. For that there was no purely scientific answer, for the quest involves a definition by people of what makes them what they are. A reply required more than looking "to the other side of the river." It meant looking across that greater rift that had grown between the moieties of formal learning. It would mean giving credence to the fossil suids and elephantids of the humanities. It would mean granting standing to narrative and to sources of understanding not grounded in natural science. It would recognize that science, too, has its context, and is

known as much by what it is not and cannot do as by what it is
and can.

Pleistocene Panorama

Even as the Great Rift pulls farther apart, responding to apparently
implacable stresses beneath the landscape of learning that have
tugged at it for a very long time, an understanding of the Pleistocene
as epoch and as idea still requires both cliffs in the same field of
vision.

The arts and humanities can no longer claim—even pretend to
claim—that they can make valid statements about the material world
and how it works. Except by casual analogy, nature cannot be considered a text and cannot be read by exegesis or hermaneutics. But
philosophy, literature, and history can help explain how the sciences
work, and they can turn the data excavated by natural science into
prisms of meaning. They can illuminate how we might understand
and express the practice of knowing, and how we come to a felt sense
of who we are and how we should behave. They can show how the
sciences themselves have evolved, full of quirks and extinct ideas;
how scientific knowledge is not revelation secularized; how science
must operate within limits or succumb to scientism.

The other cliff, its ramparts manned by the sciences, has its vantage points as well, and these train powerful but narrow lenses onto
the valley. They have, however, the weakness of their strength. They
can focus precisely on those features that their instruments can magnify, purging away the gauze of metaphysics, but they must simplify
Objekts that lie beyond their purview with artifacts that fall within
their gaze. By shearing away context as clutter, they can strip the
meaning that setting conveys. Science can analyze what caused the
scratches on a bone and what hominin skull most closely matches
others and what date the first stone tools appeared, but it cannot,
unaided, address what it means to live, what makes a life worth living, what purpose spans the narrative of a life or of humanity. It cannot say why a society should even engage in this kind of inquiry. But

science's intrinsic limits are not a cause for concern. The troubling issue comes when self-identified sciences harden into scientism and deny other inquiries a right to join in the conversation.

For now the African Rift seems to have paused, while the Renaissance rift in understanding appears destined to widen, and that which is why the boundaries of the Pleistocene will continue to shuffle and lurch into the future.

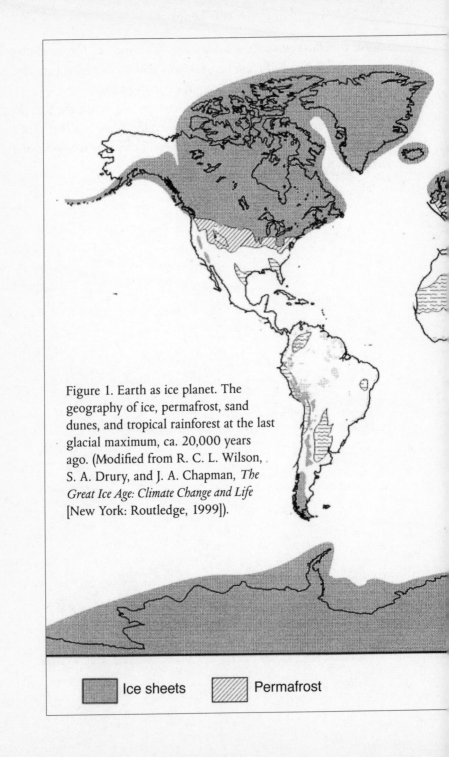

Figure 1. Earth as ice planet. The geography of ice, permafrost, sand dunes, and tropical rainforest at the last glacial maximum, ca. 20,000 years ago. (Modified from R. C. L. Wilson, S. A. Drury, and J. A. Chapman, *The Great Ice Age: Climate Change and Life* [New York: Routledge, 1999]).

Ice sheets   Permafrost

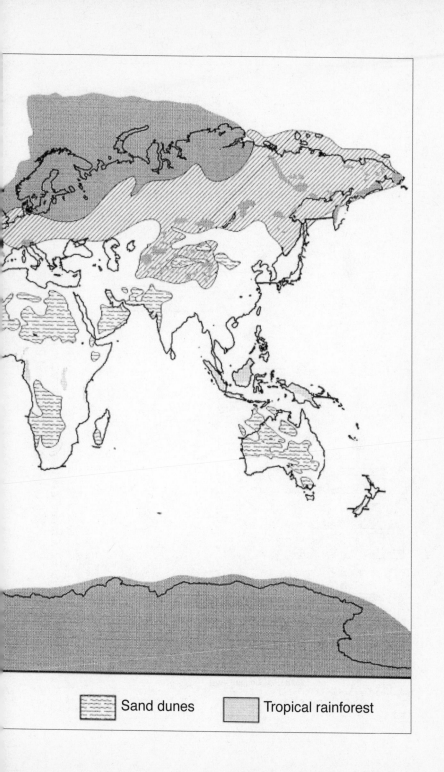

Sand dunes   Tropical rainforest

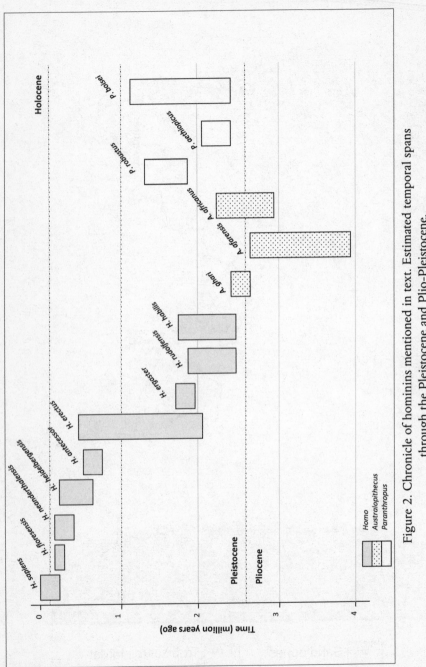

Figure 2. Chronicle of hominins mentioned in text. Estimated temporal spans through the Pleistocene and Plio-Pleistocene.

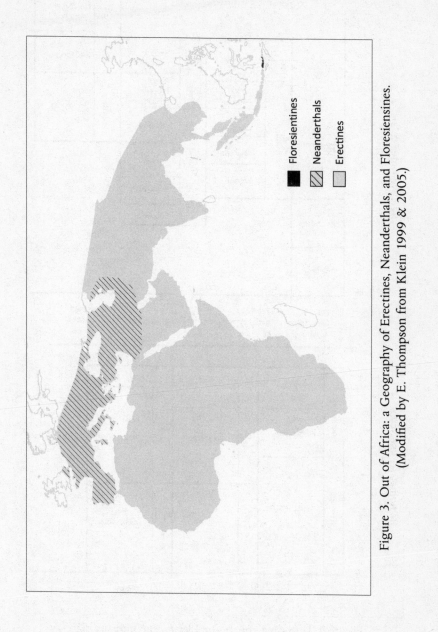

Figure 3. Out of Africa: a Geography of Erectines, Neanderthals, and Floresiensines. (Modified by E. Thompson from Klein 1999 & 2005.)

Floresientines

Neanderthals

Erectines

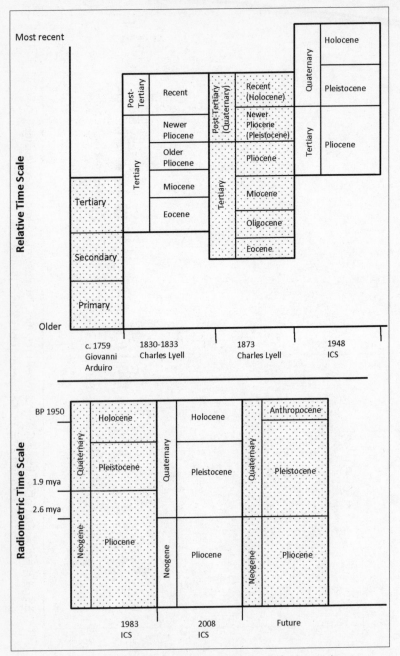

Figure 4. Unsettled epoch: the Pleistocene finds its place
in the geological time scale.

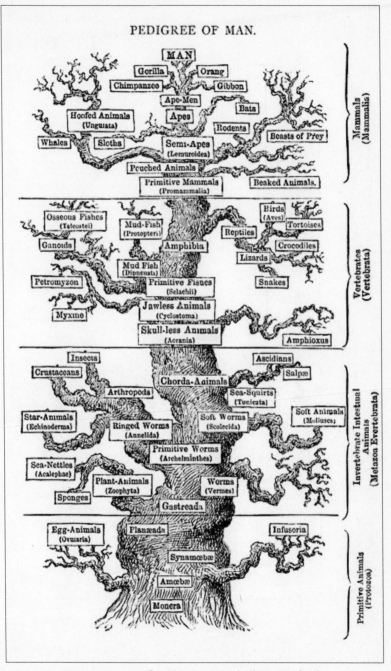

Figure 5. Icon of evolution: Haeckel's Tree of Life.

Authors' Note

This is Lydia's book. She conceived it, she supplied the information that fills it, she understood its contexts, implications, and significance. Steve served as sounding board, first-draft writer, and editor. The give-and-take between us was constant. It's Lydia's vision, Steve's voice. The idea emerged some years ago during an interglacial in our careers when we both shared an office. Left to his own devices Steve would likely have followed the spoor of fire on Earth and maybe the further exploration of planetary moons. Instead we found a new world in that most recent of past ones.

A note on titles. We played with "The Great Game" for Part 2 and liked it. With some fussing we reworked other Kipling titles to suit Parts 1 and 3. There was no special meaning intended, only an authorial whimsy, trimmed into a general consistency. The book's title was invented during our original banter, and when we began serious research, we found that it had also served as an opening chapter title in Paul Martin's *The Twilight of the Mammoths*. We are happy here to acknowledge his original introduction of the phrase to Pleistocene studies. Perhaps that world is not so lost as we imagined.

We owe debts to many others who have helped us at various stages. Thanks to Ben Minteer, Caley Orr, and Julien Riel-Salvatore, for comments on portions of drafts (some of which were more crude than we believed), and to Erin Thompson for help in converting sketches into publishable figures. To Geri Thoma for staying with a confused early draft. To Wendy Wolf for shrewd yet gentle queries that allowed us to realize early for ourselves (and to correct) what harsher critics would otherwise have had a chance to say much later.

Notes

Prologue: Mossel Bay, South Africa

1. Paul S. Martin, *The Twilight of the Mammoths* (Berkeley: University of California Press, 2007), phrase used as title to chapter 1.

Chapter 1: Rift

1. An excellent summary of the Rift's vast literature is Celia K. Nyamweru, "The African Rift System," in *The Physical Geography of Africa*, eds. William M. Adams et al. (New York: Oxford University Press, 1996), pp. 18–33. The actual timing, as always, is elusive. For a more contemporary estimate (and one linked to hominin evolution), see M. Royhan Gani and Nahid D. S. Gani, "Tectonic Hypothesis of Human Evolution," *Geotimes* (January 2008).
2. William James, "The Present Dilemma in Philosophy," *The Writings of William James,* ed. John M. McDermott (New York: Modern Library, 1967), pp. 366, 367.
3. Ibid., p. 362.

Chapter 2: Ice

1. Louis Agassiz, *Studies on Glaciers, preceded by the Discourse of Neuchâtel*, trans. and ed. Albert V. Carozzi (New York: Hafner, 1967), p. 314.
2. Herbert Butterfield, *The Origins of Modern Science* (New York: Free Press, 1997), p. 192.
3. John Imbrie and Katherine Imbrie, *Ice Ages* (Cambridge: Harvard University Press, 1979), pp. 189–91; P. E. Calkin, "Global Glacial Chronologies and Causes of Glaciation," in *Modern and Past Glacial Environments*, ed. John Menzies (Oxford: Butterworth-Heineman, 2002), pp. 15–52.
4. Actual numbers are as follows: The eccentricity cycle is 95,800 years, although it may show separate periods of 95,000 and 123,000 years, and perhaps a long cycle of 412,000 years. These numbers and the Milankovitch maximum come from Raymond Bradley, *Paleoclimatology: Reconstructing Climates of the Quaternary* (San Diego: Academic Press, 1999), p. 41. An excellent distillation of the arguments is available in John Lowe and Mike Walker, *Reconstructing Quaternary Environments* (Essex, UK: Prentice Hall, 1997), pp. 12–16.

5. For a critique of simple Milankovitch cycling, see R. C. L. Wilson, S. A. Drury, and J. A. Chapman, *Great Ice Age*, pp. 149–53.

6. See Lowe and Walker, *Reconstructing Quaternary Environments*, pp. 61–68.

7. For a survey of pluvial lakes, see Lowe and Walker, *Reconstructing Quaternary Environments*, pp. 114–21. An excellent summary exists in E. C. Pielou, *After the Ice Age* (Chicago: University of Chicago Press, 1991), pp. 192–99. Data on Canadian water supplies are from *Atlas of Canada*, http://atlas.nrcan.gc.ca/site/english/maps/freshwater/1, accessed August 8, 2009. On terrestrial water and ice generally, see G. M. Ashley, "Glaciolacustrine Environments," in Menzies, *Modern and Past Glacial Environments*, pp. 335–59.

8. See R. C. L. Wilson et al., *Great Ice Age*, and Lowe and Walker, *Reconstructing Quaternary Environments*, pp. 357–60.

9. On carbon and clathrates, see R. C. L. Wilson et al., *Great Ice Age*, pp. 108–10, 135–36. A good survey of greenhouse gases and ice ages is available in William F. Ruddiman, *Plows, Plagues, and Petroleum: How Humans Took Control of Climate* (Princeton: Princeton University Press, 2005).

10. A succinct summary keyed to the Pleistocene is available in R. C. L. Wilson et al., *Great Ice Age*, pp. 174–81.

11. Herbert Butterfield, *The Origins of Modern Science* (New York: Free Press, 1997), p. 191.

12. Forbes quote from Imbrie and Imbrie, *Ice Ages*, p. 41. This book gives a very good distillation of the theory and its history.

13. There are many good accounts of how geology created these periods. For a thumbnail, see William B. N. Berry, *Growth of a Prehistoric Time Scale Based on Organic Evolution* (San Francisco: W. H. Freeman, 1968); for the Quaternary, pp. 76–78. Perhaps the most richly textured is Martin J. S. Rudwick's two works, *Bursting the Limits of Time: The Reconstruction of Geohistory in the Age of Revolution* (Chicago: University of Chicago Press, 2005) and *Worlds Before Adam: The Reconstruction of Geohistory in the Age of Reform* (Chicago: University of Chicago Press, 2008).

14. Charles Lyell, *Principles of Geology*, Vol. III (London: John Murray, 1837), p. 373. Lyell gives a good description of his naming process in *Elements of Geology* (Philadelphia: James Kay, Un and Brother, 1839), pp. 165–66.

15. Edward Forbes, "On the connection between the distribution of existing fauna and flora of the British Isles, and the geological changes which have affected their area, especially during the epoch of the Northern Drift," Great Britain, *The Memoirs of the Geological Survey of Great Britain* 1 (1846): 336–42.

16. Quote is from Rudolf Trümpy, "The 18th International Geological Congress, Great Britain, 1948," *Episodes* 27 (3): 197. A useful survey of the history of Pleistocene chronological definitions is in Lowe and Walker, *Reconstructing Quaternary Environments*, pp. 3–8, and also Imbrie and Imbrie, *Ice Ages*, p. 152.

17. The most accessible history of developments is Imbrie and Imbrie, *Ice Ages*, pp. 123–73. An interesting (and brief) review of how working geologists reconciled the new chronometers with stratigraphic realities is Paul E. Damon,

Glen A. Izett, and Charles W. Naeser, conveners, "Pliocene and Pleistocene Geochronology," Penrose Conference Report, *Geology* 4 (October 1976): 591–93. There are several texts that lay out the techniques in detail; see, e.g., Lowe and Walker, *Reconstructing Quaternary Environments*, and Bradley, *Paleoclimatology*.

18. For a brief account of the debate, see Amanda Leigh Mascarelli, "Quaternary geologists win timescale vote," *Nature* 459 (June 4, 2009): 624.

19. The quote from Brad Pillans and Tim Naish is in "Defining the Quaternary," *Quaternary Science Reviews* 23 (2004): 2272. The article provides a very useful survey of the evolving debate.

20. Wallace Stegner, "A Sense of Place," in *Where the Bluebird Sings to the Lemonade Springs* (New York: Modern Library, 2002), p. 205.

Chapter 3: Story

1. Charles Darwin, *The Descent of Man, and Selection in Relation to Sex* (Adamant Media, 2005; reprint of 1875 ed.), p. 155.

2. Ibid.

3. We follow closely the excellent summary in Andrew S. Goudie, "Climate: Past and Present," *The Physical Geography of Africa,* eds. William M. Adams et al., (New York: Oxford University Press, 1996), pp. 34–59, and although there is considerable overlap, Andrew Goudie, *Environmental Change* (New York: Oxford University Press, 1992), pp. 97–132.

4. Data from Andrew Goudie, *The Human Impact on the Natural Environment: Past, Present, and Future* (Malden, MA: Wiley-Blackwell, 2005), p. 55; on ice effects, see p. 48.

5. Goudie, *Human Impact on the Natural Environment,* pp. 48–56.

6. Ibid., pp. 57–58.

7. Elisabeth Vrba, "Environment and Evolution: Alternative Causes of the Temporal Distribution of Evolutionary Events," *South African Journal of Science* 81 (1985): 229–36.

8. Quote from Adams, "Savanna Environments," in *Physical Geography of Africa*, p. 205.

9. Vrba, "Environment and Evolution," p. 2.

10. Karl Popper, *The Logic of Scientific Discovery* (New York: Harper Torchback, 1959), pp. 278–81.

11. Thomas Kuhn, *The Structure of Scientific Revolutions* 2nd ed. (Chicago: University of Chicago, 1971), p. 2.

12. Niles Eldredge and Stephen Jay Gould, "Punctuated Equilibria: An Alternative to Phyletic Gradualism," in *Models in Paleobiology*, ed. T. J. M. Schopf (San Francisco: Freeman, Cooper, 1972), pp. 82–115; quotes from a later version: Stephen Jay Gould and Niles Eldredge, "Punctuated Equilibria: The Tempo and Mode of Evolution Reconsidered," *Paleobiology* 3 (1977): 115–51, especially p. 115.

13. See L. Van Valen, "A New Evolutionary Law," *Evolutionary Theory* 1 (1973): 1–30, and Vrba's reply, "Turnover Pulse, Red Queen, and Related Topics," *American Journal of Science* 293A (1993): 418–52.

14. Kuhn, *The Structure of Scientific Revolutions* 2nd ed., p. 88; Popper, *The Logic of Scientific Discovery,* p. 280.

15. William Shakespeare, *Julius Caesar,* Act I, Scene ii, 140–42.

16. Richard Klein, *The Human Career* 2nd ed. (Chicago: University of Chicago Press, 1999), p. 140.

17. Major papers were M. Goodman, "The Role of Immunochemical Differences in the Phyletic Development of Human Behavior," *Human Biology* 33 (1961): 131–62; and V. M. Sarich and A. C. Wilson, "Quantitative Immunochemistry and the Evolution of Primate Albumins: Micro-Complement Fixation," *Science* 154 (1966): 1563–66; "Rates of Albumin Production in Primates," *Proceedings of the National Academy of Sciences* 58 (1967): 142–48; and "Immunological Time Scale for Hominid Evolution," *Science* 158 (1967): 1200–1203.

   An excellent popular account of how the field developed is contained in Luigi Luca Cavalli-Sforza and Francesco Cavalli-Sforza, *The Great Human Diasporas: The History of Diversity and Evolution,* trans. Sarah Thorne (Boston: Addison-Wesley Publishing, 1995), which distills the more technical L. Luca Cavalli-Sforza, Paolo Menozzi, and Alberto Piazza, *The History and Geography of Human Genes,* abridged ed. (Princeton: Princeton University Press, 1994).

18. V. M. Sarich, "A Molecular Approach to the Question of Human Origins," in *Background for Man,* eds. V. M. Sarich and P. J. Dolhinow (New York: Little, Brown, 1971): 60–61, 76. Rebecca L. Cann, Mark Stoneking, and Allan C. Wilson, "Mitochrondrial DNA and Human Evolution," *Nature* 325 (January 1, 1987): 31–36. On the controversies stirred up, see Roger Lewin, "DNA Clock Conflict Continues," *Science* 241 (September 1988), pp. 1756–59; and "Afterword," *Bones of Contention* 2nd ed. (Chicago: University of Chicago Press, 1997), pp. 321–35.

19. Quote from Matt Cartmill, "Four Legs Good, Two Legs Bad: Man's Place (If Any) in Nature," *Natural History* 92 (11): 69.

20. William S. Laughlin, "Hunting: An Integrating Behavior System and Its Evolutionary Importance," p. 304; and Sherwood L. Washburn and C. S. Lancaster, "The Evolution of Hunting," pp. 299, 303, in eds. Richard B. Lee and Irven DeVore, *Man the Hunter* (Berlin: Aldine de Gruyter, 1968).

21. Cartmill, "Four Legs Good, Two Legs Bad," p. 75.

22. William James, "The Present Dilemma in Philosophy," in *Collected Works,* pp. 363–64.

23. Ibid.

24. Huxley is quoted in Karl Popper, *The Poverty of Historicism* (New York: Routledge, 2002), p. 99. For a good illustration of a naive contemporary version, see Christine S. VanPool and Todd L. VanPool, "The Scientific Nature of Postprocessualism," *American Antiquity* 64 (January 1999): 33–53.

25. See F. M. Cornford, *Before and After Socrates* (New York: Cambridge University Press, 1932). On arche, see also Cornford, *Principium Sapientiae: The Origins of Greek Philosophical Thought*, ed. W. K. C. Guthrie (New York: Harper Torchbook, 1965), p. 172.
26. See, for example, Claude Lévi-Strauss, *Myth and Meaning: Cracking the Code of Culture* (New York: Schocken Books, 1979).
27. Paul Ricoeur, "Life in Quest of Narrative," in *On Paul Ricoeur: Narrative and Interpretation*, ed. David Wood (New York: Routledge, 1991), pp. 22–23, 32.

Chapter 4: Footnotes to Plato

1. M. D. Leakey, "A Review of the Oldowan Culture from Olduvai Gorge, Tanzania," *Nature* 210 (1966): 462–66. For the original definition, see Louis Leakey, *Stone Age Africa: An Outline of Prehistory in Africa* (New York: Oxford University Press, 1936). Other good introductions to the topic: G. L. Isaac, "The Archaeology of Human Origins: Studies of the Lower Pleistocene in East Africa, 1971–1981," in *Advances in World Archaeology*, eds. F. Wendorf and A. Close (New York: Academic Press, 1981), pp. 1–87; and "The Earliest Archaeological Traces," in *Cambridge History of Africa*, ed. J. Desmond Clark (New York: Cambridge University Press, 1982), pp. 157–247; H. J. Deacon and J. Deacon, *Human Beginnings in South Africa* (Cape Town: David Philip Publishers, 1999); Nicholas Toth and Kathy Schick, "The First Million Years: The Archaeology of Protohuman Culture," in *Advances in Archaeological Method and Theory*, ed. Michael Schiffer (New York: Academic Press, 1986), pp. 1–97.
2. Richard Leakey, *The Making of Mankind* (New York: Dutton, 1981), 65–66. See also Richard Leakey and Roger Lewin, *Origins: What New Discoveries Reveal About the Emergence of Our Species and Its Possible Future* (New York: Dutton, 1977).
3. Richard Klein, *Human Career* 2nd ed. (Chicago: University of Chicago Press, 1999), p. 219.
4. Ibid., p. 228, for ages of artifacts.
5. Ibid., p. 182, for *A. garhi* and tools.
6. For information about Oldowan technology, see Klein, *Human Career* 2nd ed., pp. 228–37; the Dart reference is on p. 228. Other valuable sources: Steve Jones, Robert Martin, and David Pilbeam, eds., *The Cambridge Encyclopedia of Human Evolution* (New York: Cambridge University Press, 1994); Chris Stringer and Peter Andrews, *The Complete World of Human Evolution* (New York: Thames & Hudson, 2005); Charles Lockwood, *The Human Story: Where We Come From and How We Evolved* (New York: Sterling, 2008).
7. Klein, *Human Career* 2nd ed., p. 222.
8. Bernard Wood, "Who Is the 'Real' Homo Habilis?" *Nature* 327 (6119): 187–88.
9. Richard Leakey and Roger Lewin, *Origins Reconsidered: In Search of What Makes Us Human* (New York: Anchor Books, 1992), p. 26.

10. William Buckland, *Geology and Mineralogy*, Vol. I (London: William Pickering, 1836), pp. 83–84.

11. Don Johanson and James Shreeve, *Lucy's Child: The Discovery of a Human Ancestor* (New York: Viking, 1989), pp. 195–96.

12. Misia Landau, *Narratives of Human Evolution* (New Haven: Yale University Press, 1991).

13. Whether they are in truth two species or regional variants of one is unknown. See Wood, "Who Is the 'Real' Homo Habilis?"

14. Elisabeth S. Vrba, "Mammals as a Key to Evolutionary Theory," *Journal of Mammalogy* 73, no. 2 (February 1992): 1–28. Quote from Klein, *Human Career* 2nd ed., p. 249; Elisabeth Vrba, "The Significance of Bovid Remains as Indicators of Environment and Predation Patterns," in *Fossils in the Making*, eds. A. K. Behresmeyer and A. P. Hill (Chicago: University of Chicago Press, 1980), pp. 247–71; Vrba, "The Fossil Record of African Antelopes (Mammalia, Bovidae) in Relation to Human Evolution and Paleoclimate," in *Paleoclimate and Evolution with Emphasis on Human Origins*, eds. E. S. Vrba et al. (New Haven: Yale University Press, 1995), pp. 385–424.

15. Jonathan Kingdon, *East African Mammals IIB*, pp. 15–29; the quote is from p. 17; Nancy E. Todd and V. Louise Roth, "Origin and Radiation of the Elephantidae," in *The Proboscidea: Evolution and Paleoecology of Elephants and Their Relatives*, eds. Jeheskel Shoshani and Pascal Tassy (New York: Oxford University Press, 1996), pp. 193–202; and Jeheskel Shoshani and Pascal Tassy, "Summary, Conclusion, and a Glimpse into the Future," in *The Proboscidea: Evolution and Paleoecology of Elephants and Their Relatives*, eds. Jeheskel Shoshani and Pascal Tassy (New York: Oxford University Press, 1996), pp. 335–48.

16. Ibid.

17. The quote on superkeystone species is from Shoshani and Tassy, "Summary, Conclusions, and a Glimpse," in *Proboscidea*, eds. Shoshani and Tassy, p. 345. For a fascinating overview of the emerging moral dimension, see Christen Wemmer and Catherine A. Christen, eds., *Elephants and Ethics: Toward a Morality of Coexistence* (Baltimore: Johns Hopkins University Press, 2008).

18. David Pilbeam, "Foreword" in John Reader, *Missing Links: The Hunt for Earliest Man* (New York: Penguin Books, 1988), p. x.

19. Lionel Tiger, "My Ardi, Myself," Forbes.com, accessed on October 6, 2009.

20. The data for the 2011 Paleoanthropology Society Conference comes from the society's published abstracts. "Abstract from the Paleoanthropology Society 2011 Meeting," *Paleoanthropology* 2011: A1–A41.

Chapter 5: Out of Africa

1. The quote is from Charles Lockwood, *The Human Story* (New York: Sterling, 2008), p. 69.

2. G. Philip Rightmire, "Patterns of Hominid Evolution and Dispersal in the Middle Pleistocene," *Quaternary International* 75 (2001): 77–84; the quote is from page 82; Klein, *Human Career* 2nd ed., p. 280.

3. On the Pleistocene colonization of islands, see R. G. Bednarik, "Replicating the First Known Sea Travel by Humans: The Lower Pleistocene Crossing of Lombok Strait," *Human Evolution* 16, nos. 3–4 (2001): 229–42; the quote is from page 340. Recent discoveries on Crete have confirmed an erectine colonization.

4. The Huxley quote is in O. C. Marsh, "Thomas Henry Huxley," *American Journal of Science* L, no. 296 (August 1895): 177–83; the quote is from page 181. The Marsh quote is in "Notice of New Equine Mammals from the Tertiary Formation," *American Journal of Science* VII, no. 39 (March 1874): 258. For information about equids, we have relied extensively on Bruce J. MacFadden, *Fossil Horses* (Cambridge: Cambridge University Press, 1992), and recommend it to any investigator.

5. G. G. Simpson, *Horses: The Story of the Horse Family in the Modern World and Through Sixty Million Years* (New York: Anchor, 1961), p. 168.

6. We follow MacFadden, *Fossil Horses*, pp. 156–64. For a distilled version with an excellent phylogenic diagram, see Bruce J. MacFadden, "Fossil Horses— Evidence for Evolution," *Science* 307 (March 18, 2005): 1728–30.

7. This episode is quoted in William H. Goetzmann, *Exploration and Empire* (New York: Knopf, 1966), p. x; the MacFadden quote is from MacFadden, *Fossil Horses*, p. 1.

8. This section follows closely the article by R. W. Wrangham and N. L. Conklin-Brittain, "The Biologicial Significance of Cooking in Human Evolution," *Comparative Biochemistry and Physiology* Part A 136 (2003): 35–46. These ideas have been elaborated (though not enhanced) in Richard Wrangham, *Catching Fire: How Cooking Made Us Human* (New York: Basic Books, 2009); see especially pages 55 to 81.

9. Wrangham, *Catching Fire*, p. 18 (he quotes a study done in Germany). On myths, see Sir James Frazer, *Myths on the Origin of Fire* (New York: Macmillan, 1930; reprint, New York: Hacker Art Books, 1974).

10. Wrangham, *Catching Fire*, p. 40. See also Johan Goudsblom, *Fire and Civilization* (New York: Penguin, 1992), pp. 33–37.

11. Wrangham, *Catching Fire*, p. 98.

12. From "Fire Out of Africa" a news release of the Hebrew University of Jerusalem (October 27, 2008); www.huji.ac.il/cgi-bin/dovrut/dovrut_search_eng .pl?mesge122510374832688760, accessed on October 24, 2009. Other sources for information on early fire: On the oldest likely hearth, J. A. J. Gowlett et al., "Archaeological Sites, Hominid Remains and Traces of Fire from Chesowanja, Kenya," *Nature* 294 (November 23, 1981): 125–29, and in an exchange of comments, *Nature* 296 (April 29, 1982): 870; Randy V. Bellomo, "Methods of Determining Early Hominid Behavioral Activities Associated with the Controlled Use of Fire at FxJj 20 Main, Koobi Fora, Kenya," *Journal of Human Evolution* 27 (1994): 173–95.

13. The Pliny quote is from Cyril Stanley Smith and Martha Teach Gnudi, trans. and eds., *The Pirotechnia of Vannoccio Biringuccio* (Cambridge: MA: MIT Press, 1966; reprint), p. 336; the Boerhaave quote is from Gaston Bachelard, *The Psychoanalysis of Fire* (Boston: Beacon Press, 1964), p. 60.

14. Lydia Pyne and Manfred Laubichler, "Niche Construction Theory: Archaeology and Explanation," Society for American Archaeology Conference 2008, organized session.

15. Carl Sauer, "Fire and Early Man," in *Land and Life: A Selection from the Writings of Carl Ortwin Sauer*, ed. John Leighly (Berkeley: University of California Press, 1963), p. 295; Loren Eiseley, "Man the Firemaker," in *The Star Thrower* (New York: Harcourt Brace Jovanovich, 1978), pp. 47, 49; Pierre Teilhard de Chardin, *The Phenomenon of Man* (New York: Harper Torchback, 1976), p. 160; Claude Lévi-Strauss, *The Raw and the Cooked: Mythologiques*, Vol. 1, p. 164; Leach quote is from Wrangham, *Catching Fire*, p. 12.

16. See Thomas Glick, *The Comparative Reception of Darwin* (Chicago: University of Chicago Press, 1988).

17. Harvey M. Feinberg and Joseph B. Solodow, "Out of Africa," *Journal of African History* 43 (2002): 255–61; the quote is from page 255. The substance of our text derives from this delightful essay, with some tweaking to fit our general metaphorical context.

18. Ibid., p. 258.

19. Ibid., pp. 259–60.

20. Ibid., pp. 256–57, 259–60.

21. Ibid., pp. 255–56. The authors offer no explanation for how Dinesen's title was changed but suggest, a bit lamely, that inquiries into the Karen Blixen archives at the Royal Library in Copenhagen might help.

22. Ibid., p. 25.

Chapter 6: Missing Links

1. E. M. W. Tillyard, *The Elizabethan World Picture* (New York: Vintage Books, 1959), pp. 25-26.

2. Leibniz is quoted in A. O. Lovejoy, *The Great Chain of Being* (Cambridge, MA: Harvard University Press, 1964), p. 149.

3. Ibid., p. vii. Anyone at all familiar with this topic will recognize our debt to Lovejoy's indispensable monograph.

4. Tillyard, *Elizabethan World Picture*, p. 2.

5. Lovejoy, *Great Chain of Being*, p. 59; William Shakespeare, *The Tempest*.

6. Alexander Pope, *Essay on Man*, line 267.

7. Tillyard, *Elizabethan World Picture*, pp. 27–28.

8. On Leibniz, see Lovejoy, *Great Chain of Being*, p. 145, and the quote from Locke is on page 184.

9. C. Northcote Parkinson, *Parkinson's Law: Or the Pursuit of Progress* (New York: Penguin, 2002), p. 60.

10. Lovejoy, *Great Chain of Being*, p. 183; Parkinson, *Parkinson's Law*, p. 61.

11. Lovejoy, *Great Chain of Being*, p. 183.

12. William H. Goetzmann, *New Lands, New Men: America and the Second Great Age of Discovery* (New York: Viking, 1986).

13. Stephen Toulmin, *The Discovery of Time* (London: Octagon Books, 1983).

14. Pope, *Essay on Man*, lines 244–46.

15. Ernst Haeckel, *The Last Link: Our Present Knowledge of the Descent of Man* (1899), p. 3.

16. Ernst Haeckel, *The Evolution of Man*, Vol. 2 (New York: D. Appleton, 1886), pp. 294, 326–27.

17. Haeckel, *Last Link*, p. 76.

18. We follow closely the account in John Reader, *Missing Links* (New York: Penguin Press Science, 1989), p. 7.

19. Again, we follow Reader, *Missing Links*, pp. 10–15.

20. See various editions of *Systema Naturae*; for their biographical setting, see the delightful Wilfrid Blunt, *Linnaeus: The Compleat Naturalist* (Princeton: Princeton University Press, 2001). For quote, see a letter to Johann Georg Gmelin dated February 25, 1747.

21. Caroli Linnaei, *Systema Naturae* (Holmiae: Laurantii Salvii, 1758) vol. 1, p 24.

22. Haeckel, *Last Link*, pp. 12–13.

23. Our account follows the study by John Wolf and James S. Mellett, "The Role of 'Nebraska Man' in the Creation/Evolution Debate," *Creation/Evolution* 16 (1985): 31–43.

24. Ibid.

25. Ibid.

26. A robust library of books and articles exists regarding Piltdown. We have relied primarily on Charles Blinderman, *The Piltdown Inquest* (Buffalo, NY: Prometheus Books, 1986) and Frank Spencer, *Piltdown: A Scientific Forgery* (Oxford University Press, 1990). See also John E. Walsh, *Unraveling Piltdown: The Science Fraud of the Century and Its Solution* (New York: Random House, 1996), and for a handy distillation, the chapter in Reader, *Missing Links*.

27. Philip Sclater, "The Mammals of Madagascar," *Quarterly Journal of Science* (1864): 212–19; Haeckel, *The Evolution of Man* Vol. 2, trans. Joseph McCabe (London: Watts & Co., 1905), p. 635; see also. Ernst Haeckel and Edwin Ray Lankester, *The History of Creation*, Vol. 2 (New York: D. Appleton, 1876) pp. 375–76.

28. George Gaylord Simpson, "Mammals and Land Bridges," *Journal of the Washington Academy of Sciences* 30, no. 4 (1940), pp. 137–63.

29. Misia Landau, *Narratives of Human Evolution* (New Haven: Yale University Press, 1991), p. 3.

30. E. M. Forster, *Aspects of the Novel* (Harcourt, Brace, and World, 1954); Chekhov (and useful commentary) in John Gardner, *The Art of Fiction: Notes on Craft for Young Writers* (New York: Knopf, 1984).

31. Landau, *Narratives of Human Evolution*, pp. 3–16. For another take on the hero tale, see Joseph Campbell, *The Hero with a Thousand Faces* (New York: Pantheon, 1949), which expands the range of consideration.

32. Raymond Dart, "The Predatory Transition from Ape to Man," *International Anthropology and Linguistic Review* 1, no. 4 (1953): 207.

33. Karl Popper, *The Poverty of Historicism* (Routledge, 1957; reprint 2002), pp. xi, 3.

34. Stephen J. Gould, *Ever Since Darwin: Reflections in Natural History* (New York: W. W. Norton, 1977), p. 61.

35. For a good summary of these developments, see Donald K. Grayson, "Nineteenth-Century Explanations of Pleistocene Extinctions: A Review and Analysis," in *Quaternary Extinctions: A Prehistoric Revolution*, eds. Paul S. Martin and Richard G. Klein (Tucson: University of Arizona Press, 1984), pp. 5–9. Thomas Jefferson, "A memoir on the discovery of certain bones of a quadruped of the clawed kind in the western part of Virginia," *Transactions of the American Philosophical Society* 4 (1799): 246–60; the quote is on page 255.

36. Jefferson, "A memoir on the discovery of certain bones," pp. 255–56.

37. Paul Martin, *Twilight of the Mammoths* (Berkeley: University of California Press, 2005), p. 200.

38. Sergey A. Zimov, "Pleistocene Park: Return of the Mammoths Ecosystem," *Science* 308, no. 5723 (May 6, 2005): 796–98; Stefan Lovgren, "Pleistocene Park Underway: Home for Reborn Mammoths?" *National Geographic News* (May 17, 2005); http://news.nationalgeographic.com/news/pf/82726080 .html; accessed on November 2, 2009.

Chapter 7: New Truths, Heresies, Superstitions

1. Snorri Sturluson, *Gylfaginning*, trans. Arthur G. Brodeur (1916, 1923).

2. The best summary on this is Björn Kurtén, *Pleistocene Mammals of Europe* (Piscataway, NJ: Aldine Transaction, 1968).

3. The best summary on this is in Richard Klein, *The Human Career* 2nd ed. (Chicago: University of Chicago Press, 1999), pp. 435–614. There are good digests in Charles Lockwood, *The Human Story* (New York: Sterling, 2008), pp. 88–97; R. C. L. Wilson, S. A. Drury, and J. L. Chapman, *The Great Ice Age*, pp. 204–7; Ian Tattersall, *The Last Neanderthal: The Rise, Success, and Mysterious Extinction of Our Closest Human Relatives* (Boulder, CO: Westview Press, 1999).

4. Thomas H. Huxley, "Further Remarks upon the Human Remains from the Neanderthal," *Natural History Review; Scientific Memoirs II* (1864), p. 574.

5. Kurtén, *Pleistocene Mammals*, p. 119.

6. See, for example, Katherine Harmon, "Climate Change Likely Caused Polar Bear to Evolve Quickly," *Scientific American* (March 10, 2010); www.scientific american.com/article.cfm?id=polar-bear-genome-climate.

7. Quoted phrase from Albert Gaudry from Björn Kurtén, *The Cave Bear Story: Life and Death of a Vanished Animal* (New York: Columbia University Press, 1976), pp. 18, 109–24. Our account of the cave bear derives essentially from this delightful book.

8. Ibid., pp. 131–32.

9. See, for example: Editors, "Did Neanderthals Think Like Us?" *Scientific American* (June 2010): 72–75. Interest is inexhaustible.

10. R. S. Sommer and A. Nadachowski, "Glacial Refugia of Mammals in Europe: Evidence from Fossil Records," *Mammal Review* 36, no. 4 (2006): 251–65;

Godfrey M. Hewitt, "Post-glacial Re-colonization of European Biota," *Biological Journal of the Linnean Society* 68, no. 1–2 (1999): 87–112; Godfrey Hewitt, "The Genetic Legacy of the Quaternry Ice Ages," *Nature* 405 (June 22, 2000): 907–13.

11. Hewitt, "Post-glacial Re-recolonization of European Biota," pp. 87–112.

12. Paul Mellars, "A New Radiocarbon Revolution and the Dispersal of Modern Humans in Eurasia," *Nature* 439 (February 23, 2006): 931–35, especially the dispersal map on page 933.

13. Tattersall, *The Last Neanderthal*, p. 202.

14. The literature is rich, and constantly expanding, but the core reference remains Paul S. Martin and Richard G. Klein, eds., *Quaternary Extinctions: A Prehistoric Revolution* (Tucson: University of Arizona Press, 1984).

15. See, in particular Elaine Anderson, "Who's Who in the Pleistocene: A Mammalian Bestiary," in *Quaternary Extinctions*, eds. Martin and Klein, pp. 40–89; and Paul S. Martin, "Prehistoric Overkill: The Global Model," in *Quaternary Extinctions*, eds. Martin and Klein, pp. 354–403.

16. Martin, "Prehistoric Overkill."

17. Alfred W. Crosby, *Ecological Imperialism: The Biological Expansion of Europe, 900–1900* (Cambridge: Cambridge University Press, 1986), p. 80.

18. Tattersall, *The Last Neanderthal*, p. 198.

19. A modern retelling is available in John Reader, *Missing Links* (New York: Penguin Press Science, 1989), pp. 9–15. A detailed summary account from the era itself is available in Charles Lyell, *The Geological Evidences of the Antiquity of Man* (London: John Murray, 1863).

20. Lyell, *Geological Evidences,* pp. 92, 472–73, 506.

21. Thomas H. Huxley, *Man's Place in Nature* (Ann Arbor: University of Michigan Press, 1959; reprint of 1863 edition), pp. 181–82; T. H. Huxley, "Further Remarks upon the Human Remains from the Neanderthal," *Natural History Review* 4 (1864), p. 429.

22. Alfred Wallace, "The Origin of Human Races and the Antiquity of Man Deduced from the Theory of 'Natural Selection,'" *Journal of the Anthropological Society of London* 2 (1964): clxvii–clxviii.

23. Ibid.

24. Ibid., p. clxix.

25. For a composite history, see "Editorial," *The Linnean* 15, no. 2 (1999), pp. 1–13. For Busk's quotes, see George Busk, "Pithecoid Priscan Man from Gibraltar," *The Reader* (1864), p. 110.

26. David Brill, "Neanderthal's Last Stand," *Nature News* (September 13, 2006): doi: 10.1038/news060911-8; Tattersall, *Last Neanderthal,* pp. 198–203; Cecilio Barroso Ruiz and Jean Jacques Hublin, "The Late Neanderthal Site of Zafarraya (Andalucia, Spain), Gibraltar during the Quaternary," *AEQUA Monografías* 2 (1994): 61–70.

27. T. H. Huxley, *Man's Place in Nature*, p. 9.

28. Quote from Kurtén, *The Cave Bear Story*, p. 1.

29. William J. Broad, *The Oracle: Ancient Delphi and the Science Behind Its Lost Secrets* (New York: The Penguin Press, 2006).

30. Marta Mirazon Lahr and Robert Foley, "Human Evolution Writ Small," *Nature* 431 (October 28, 2004): 1043.

31. P. Brown et al., "A New Small-Bodied Hominin from the Late Pleistocene of Flores, Indonesia," *Nature* 431 (October 28, 2004): 1055–61; M. J. Morwood, "Archaeology and Age of a New Hominin from Flores in Eastern Indonesia," *Nature* 431 (October 28, 2004): 1087–91; Rex Dalton, "Little Lady of Flores Forces Rethink of Human Evolution," *Nature* 431 (October 28, 2004): 1029. Other background sources: Chris Stringer, "A Stranger from Flores," *Naturenews* (October 27, 2004): doi:10.1038/news041025-3; Klein, *Human Career* 3rd ed. For a lively but self-serving critique of the "new species" argument by its most prominent dissenters, see Maciej Henneberg, Robert B. Eckhardt, and John Schofield, *The Hobbit Trap*, 2nd ed. (Walnut Creek, CA: Left Coast Press, 2011).

32. The quotes are from Lahr and Foley, "Human Evolution Writ Small," p. 1043; Michael Hopkin, "The Flores Find," *Naturenews* (October 27, 2004): doi:10.1038/news041025-4.

33. Klein, *The Human Career* 3rd ed., p. 722.

34. Gregory Forth, "Hominids, Hairy Hominoids and the Science of Humanity," *Anthropology Today* 21, no. 3 (June 9, 2005): 13–17.

35. Dalton, "Little Lady of Flores," p. 1029; Brown et al., "A New Small-Bodied Hominin."

36. Quote from Dalton, "Little Lady of Flores," p. 1029.

37. Klein, *The Human Career* 3rd ed., p. 724.

38. Richard Roberts, "Villagers Speak of the Small, Hairy Ebu Gogo," *Daily Telegraph* (October 28, 2004).

39. See Gregory Forth, "Hominids, Hairy Hominoids and the Science of Humanity," *Anthropology Today* 21, no. 3 (June 9, 2005): 13–17; Editor, "The People Time Forgot: Flores Find," *National Geographic* (April 2005).

40. Claude Lévi-Strauss, "When Myth Becomes History," *Myth and Meaning: Cracking the Code of Culture* (Chicago: University of Chicago Press, 1975), pp. 34–35, 43.

Chapter 8: The Ancients and the Moderns

1. Richard Klein, *The Human Career* 2nd ed. (Chicago: University of Chicago Press, 1999), pp. 615–724; Charles Lockwood, *The Human Story* (New York: Sterling, 2008), pp. 100–107.

2. Stanley H. Ambrose, "Late Pleistocene Human Population Bottlenecks, Volcanic Winter, and Differentiation of Modern Humans," *Journal of Human Evolution* 34 (1998): 623–51; M. R. Rampino and S. Self, "Climate-Volcanism Feedback and the Toba Eruption," *Quaternary Research* 40 (1993): 269–80.

3. Ambrose, "Late Pleistocene Human Population Bottlenecks," op. cit.

4. This is the best expression of competing viewpoints: for the revolutionaries, Klein, *Human Career* 2nd ed.; and for the contras, Sally McBrearty and Alison S. Brooks, "The Revolution That Wasn't: A New Interpretation of the Origin of Modern Human Behavior," *Journal of Human Evolution* 39 (2000): 453–563.

5. Klein, *Human Career* 2nd ed., p. 492.

6. Christopher S. Henshilwood et al., "Emergence of Modern Human Behavior: Middle Stone Age Engravings from South Africa," *Science* 295 (February 15, 2002): 1279.

7. Klein, *Human Career* 2nd ed., p. 514.

8. McBrearty and Brooks, "The Revolution That Wasn't," p. 453.

9. Ibid., p. 534.

10. J. E. Spingarn, ed., *Sir William Temple's Essays on Ancient & Modern Learning and On Poetry* (Oxford: Clarendon Press, 1909), pp. 2, 20.

11. Jonathan Swift, *Gulliver's Travels, A Tale of a Tub, Battle of the Books, Etc.*, ed. William Alfred Eddy (New York: Oxford University Press, 1933), p. 544.

12. Ibid.

13. Karl Popper, "Natural Selection and the Emergence of Mind," *Dialectica* 32 (1978), p. 348.

14. Ibid., p. 351.

15. Marta Mirazon Lahr and Robert Foley, "Human Evolution Writ Small," *Nature* 431 (October 28, 2004): 1044.

16. Francesco d'Errico, Christopher Henshilwood, and Peter Nilssen, "An Engraved Bone Fragment from c. 70,000-Year-Old Middle Stone Age Levels at Blombos Cave, South Africa: Implications for the Origin of Symbolism and Language," *Antiquity* 75 (2001): 317.

17. This quote is the opening line of d'Errico et al., "An Engraved Bone Fragment," p. 309.

18. Swift, *Gulliver's Travels*, p. 21.

19. Examples cited in Misia Landau, *Narratives of Human Evolution* (New Haven: Yale University Press, 1993), pp. 165–67.

20. Karl Popper, "Natural Selection and the Emergence of Mind," pp. 342, 352.

21. Quoted in James Graff, "Saving Beauty," *Time* (May 29, 2002). See also Graff, "The Lessons of Lascaux," *Time* (May 7, 2006).

22. Bataille, *Lascaux*, p. 11. The Picasso quote is from Albert Skira in George Bataille, *Lascaux, or The Birth of Art* (Skira, n.d.), preface.

23. Ibid., p. 11.

24. Molly Moore, "Debate Over Moldy Cave Art Is a Tale of Human Missteps," *Washington Post* (July 1, 2008). On Altamira, see Graff, "The Lessons of Lascaux."

Chapter 9: The Hominin Who Would Be King

1. The principal source for this section is Paul S. Martin and Richard G. Klein, eds., *Quaternary Extinctions: A Prehistoric Revolution* (Tucson: University of

Arizona Press, 1984). Many studies have since debated the causes; for a sampling of the recent literature, see A. D. Barnosky et al., "Assessing the Causes of Late Pleistocene Extinctions on the Continents," *Science* 306 (2004): 70–75; and P. L. Koch and A. D. Barnosky, "Late Quaternary Extinctions: State of the Debate," *Annual Review of Ecology and Evolutionary Systems* 37 (2006): 215–50. Estimated rate of loss from Peter M. Vitousek et al., "Human Domination of Earth's Ecosystems," *Science* 277 (July 25, 1997): 498.

2. For an excellent summary, see Juliet Clutton-Brock, *A Natural History of Domesticated Mammals* (New York: Cambridge University Press, 1999).

3. Anthony D. Barnosky, "Megafauna Biomass Tradeoff as a Driver of Quaternary and Future Extinctions," Proceedings National Academy of Science 105, Suppl. 1 (August 12, 2008): 11543–48; and Vaclav Smil, *The Earth's Biosphere: Evolution, Dynamics, Change* (Cambridge, MA: MIT Press, 2003), p. 284.

4. R. C. L. Wilson, S. A. Drury, J. L. Chapman, *The Great Ice Age* (New York: Routledge, 2000), p. 235.

5. The best summary is Stephen J. Pyne, *Fire: A Brief History* (Seattle: University of Washington Press, 2001).

6. For a good statistical distillation, see Will Steffen, Paul J. Crutzen, and John R. McNeill, "The Anthropocene: Are Humans Now Overwhelming the Great Forces of Nature?," *Ambio* 36, no. 8 (December 2007): 614–21.

7. William F. Ruddiman, *Plows, Plagues, and Petroleum: How Humans Took Control of Climate* (Princeton: Princeton University Press, 2005), p. 105; Wilson et al., *Great Ice Age*, pp. 150, 156.

8. We follow the reasoning of Ruddiman, *Plows, Plagues, and Petroleum*. For a popular survey of climatic influences on human settlement prior to the Anthropocene, and how stability aided, see Brian Fagan, *The Long Summer: How Climate Changed Civilization* (New York: Basic Books, 2004); the quoted phrase is from the title.

9. See Ruddiman, *Plows, Plagues, and Petroleum*.

10. For a concise, popular summary, see Luigi Luca Cavalli-Sforza and Francesco Cavalli-Sforza, *The Great Human Diasporas: The History of Diversity and Evolution* (Boston: Addison-Wesley, 1995).

11. A useful summary of the genome is Cavalli-Sforza and Cavalli-Sforza, *The Great Human Diasporas*.

12. For a brief introduction, see Leslie A. Pray, "Epigenetics: Genome, Meet Your Environment," *The Scientist* 18 (13/14) (July 5, 2004), and Stephan Beck, Alexander Olek, and Jorn Walter, "From Genomics to Epigenomics: A Loftier View of Life," *Nature Biotechnology* 17 (December 1999): 1144; Eva Jablonka and Marion J. Lamb, *Evolution in Four Dimensions: Genetic, Epigenetic, Behavioral, and Symbolic Variation in the History of Life* (Cambridge, MA: MIT Press, 2006).

13. Smil, *The Earth's Biosphere*, p. 271.

14. On thinking like a machine, see Ken Auletta, *Googled: The End of the World as We Know It* (New York: Penguin Press, 2009), p. 328.

15. Bill McKibben, *The End of Nature* (New York: Doubleday, 1989), p. 83.
16. On Pleistocene parks, see Emma Marris, "Reflecting the Past," *Nature* 462 (November 5, 2009): 30–32; and Josh Donlan, "Re-wilding North America," *Nature* 436 (August 18, 2005): 913–14.
17. For a moderately complete survey of such themes, see Paul Shepard, *Coming Home to the Pleistocene* (Washington, DC: Island Press, 1998).

Chapter 10: The Anthropocene

1. Paul Crutzen and E. F. Stoermer, "The 'Anthropocene,'" *Global Change* 41 (2000): 17–18.
2. Faust quote from Bayard Taylor translation, Johann Wolfgang von Goethe, *Faust: A Tragedy* (Leipzig: F. A. Brockhaus, 1872), p. 9.
3. Two good summaries of the case are: Will Steffen, Paul J. Crutzen, and John R. McNeill, "The Anthropocene: Are Humans Now Overwhelming the Great Forces of Nature?," *Ambio* 36, no. 8 (December 2007): 614–21; and Jan Zaslasiewicz et al., "Are We Now Living in the Anthropocene?," *GSA Today* 18, no. 2 (February 2008): 4–8.
4. Steffen, Crutzen, McNeill, "The Anthropocene," p. 618.
5. See Zalasiewicz et al., "Are We Now Living in the Anthropocene?," pp. 6, 7, for an interesting commentary and source of quote.
6. Steffen, Crutzen, McNeill, "The Anthropocene," p. 618.
7. See Zalasiewicz et al., "Are We Now Living in the Anthropocene?," p. 4.
8. Quote from *Encyclopedia Britannica* 1998, vol. 19 (Chicago: Encyclopedia Britannica), p. 868.
9. See, for example, Paul Ricoeur, "History and Rhetoric," *Diogenes* 168 (42/4) (Winter 1994): 7–24.
10. Paul Crutzen, "Can We Survive the 'Anthropocene' Period?" Welt Online (June 16, 2009); www.welt.de/international/article3933164/Can-we-survive -the-anthropocene-period.html.
11. Interestingly, Crutzen later coauthored an essay on the Anthropocene for a more general audience in which he enlisted the assistance of John McNeill, an environmental historian best known for his study of fossil-fuel civilization, *Something New Under the Sun: An Environmental History of the Twentieth-Century World* (New York: Norton, 2001).
12. Carl L. Becker, *The Heavenly City of the Eighteenth-Century Philosophers* (New Haven: Yale University Press, 1932), p. 31.
13. Ernst Cassirer, *The Logic of the Cultural Sciences: Five Studies* (New Haven: Yale University Press, 2000), pp. 35–37.
14. Ibid.
15. The quotes are from Ernst Cassirer, *An Essay on Man* (New Haven: Yale University Press, 1944), pp. 207, 195.
16. Karl Popper, *The Poverty of Historicism* (New York: Routledge, 2002), pp. 98–99.

17. See Michael Friedman, *A Parting of the Ways: Carnap, Cassirer, and Heidegger* (Peru, IL: Open Court, 2000).
18. The quote is from Lewis R. Binford and Jeremy A. Sabloff, "Paradigms, Systematics, and Archaeology," *Journal of Anthropological Research* 38, no. 2 (Summer 1982): 139.
19. Ibid., pp. 150, 151.
20. For an interesting survey of persistence in explanation, see Wiktor Stoczkowski, *Explaining Human Origins: Myth, Imagination, and Conjecture*, trans. Mary Turton (New York: Cambridge University Press, 2002).
21. Paul Feyerabend, *Against Method* (Brooklyn: Verso, 1993), pp. 305–6.

Epilogue: Rift Redux

1. Robert Ardrey, *African Genesis: A Personal Investigation into the Animal Origins and Nature of Man* (New York: Atheneum, 1961), pp. 9, 27.
2. Ardrey, *African Genesis*, pp. 13, 184.
3. Roger Lewin, *Bones of Contention* 2nd ed. (Chicago: University of Chicago Press, 1997), chaps. 9–10. A thumbnail sketch of the geochronological literature is given in R. L. Hay, "The KBS Tuff Controversy May Be Ended," *Nature* 284 (April 2, 1980): 401.
4. Lewin, *Bones of Contention* 2nd ed., p. 189. Lewin is our primary source of information here, supplemented by select publications in the scientific literature. A detailed survey of regional studies, especially helpful for documenting the long prelude to the KBS tuff incident, is available in John M. Harris, Meave G. Leakey, and Francis H. Brown, "A Brief History of Research at Koobi Fora, Northern Kenya," *Ethnohistory* 53, no. 1 (Spring 2006): 35–67.
5. Lewin, *Bones of Contention* 2nd ed., p. 189.
6. Ibid., p. 190.
7. Ibid.
8. Ibid.

Index